Process Mixing

Chemical and Biochemical Applications: Part II

A publication of the North American Mixing Forum

Elmer L. Gaden Jr., series editor
Gary B. Tatterson, volume editor
Richard V. Calabrese and W. Roy Penney, volume co-editors

N. Ahmed
R.J. Allgor
P.M. Armenante
K. Baker
A. Bakker
G. Baldi
A.A. Barresi
O. Chikao
D.S. Dickey
F.J. Dickin
R.B. Edwards
S-O. Enfors
J.B. Fasano
R.O. Fox
E. Galindo
S. George
M.R. Grier
R.A. Howk
K. Ihara
O.M. Ilyas
K.S. Lally
G. Larsson
T. Li
F.H. Ling
K. Lodaya
I. Manas-Zloczower
R. Mann
B. Mazzarotta
J.W. Mitchell
J.Y. Murthy
K.J. Myers
H. Nakamoto
E.B. Nauman
J.Y. Oldshue
R. Parthasarathy
W.R. Penney
C.Y. Perng
M. Pipino
V.V. Ranade
E.P.L. Roberts
B. Roman
A. Rushton
M. Salvador
M.E. Sensel
R.N. Sharma
G.B. Tatterson
R.A. Williams
B. Xu
P. Ying
A.J. Zehnder

AIChE Staff
Maura Mullen, Managing Editor
Thomas E.H. Campbell, Editorial Assistant
Cover Design: Mark J. Montgomery

Inquiries regarding the publication of Symposium Series Volumes should be directed to
Dr. Elmer L. Gaden Jr., series editor
University of Virginia, Department of Chemical Engineering
Thornton Hall, Charlottesville, Virginia 22903-2442 FAX: (804) 924-6270

AIChE Symposium Series
1993

Number 293 | Volume 89

Published by
American Institute of Chemical Engineers

345 E 47 Street | New York, N.Y. 10017

American Institute of Chemical Engineers
345 East 47 Street, New York, N.Y. 10017

Library of Congress Cataloging-in-Publication Data

Process mixing
(AIChE Symposium Series ; no. 286, v. 88)
"First official publication of the North American
Mixing Forum"--Foreword
Pt. 2 edited by Gary B. Tatterson, Richard V.
Calabrese, and W. Roy Penney
Includes indexes.
1. Mixing--Congresses. I. Tatterson, Gary B.
II. Calabrese, Richard, 1946- III. Penney, W. Roy.
IV. American Institute of Chemical Engineers.
V. AIChE symposium series; no. 286.
TP156.M5P76 1992 660'.284292 92-8353
ISBN 0-8169-0611-4 (pt. 2)

FOREWORD

Mixing is part of the infrastructure of the chemical, petrochemical and biochemical industries. These industries, however, are in need of better characterization of mixing and multiphase processing because of the economic significance of mixing and contacting. It is with this purpose that this symposium was developed.

This publication is sponsored by the North American Mixing Forum.

Gary B. Tatterson, *volume editor*
Department of Chemical Engineering
North Carolina A&T State University
Greensboro, NC 27411

Richard V. Calabrese, *volume co-editor*
Department of Chemical Engineering
University of Maryland
College Park, MD 20742

W. Roy Penney, *volume co-editor*
Department of Chemical Engineering
University of Arkansas
Fayetteville, AR 72701

Organizing Committee

Richard V. Calabrese
Deepak Doraiswamy
Julian Fasano
Costas G. Gogos
Ramesh R. Hemrajani
Andrew Hrymak
Douglas E. Leng
Sanjeev Katti
Kevin Myers
Dennis Olander
Paul Szabo
Lawrence L. Tavlarides
Philip E. Wood

Officers of the North American Mixing Forum

President: James Y. Oldshue
Vice President: E. Bruce Nauman
Treasurer: Richard V. Calabrese
Secretary: Piero M. Armenante

Contents

Feedpipe Backmixing in Agitated Vessels

Julian B. Fasano

Chemineer, Inc., 5870 Poe Ave., Dayton, OH 45414

W. Roy Penney and Bangcheng Xu

Dept. of Chem. Eng., University of Arkansas, Fayetteville, AR 72701

Feedpipe backmixing has been determined experimentally for a 0.009525 m (3/8 inch) feedpipe using 6-Blade DISK and HE-3 impellers in a 0.2413 m (9.5 inch) diameter by 0.2413 m tall, liquid-filled, fully-baffled vessel using water as the test fluid. The minimum allowable feedpipe velocity to prevent feedpipe backmixing was determined as a function of impeller style, impeller speed and feedpipe location. A visual technique using dye injection was used to determine the onset of feedpipe backmixing. For a given impeller style and feedpipe location the ratio of minimum allowable feedpipe velocity to agitator tip speed (v_f/v_t) was nearly a constant. The minimum values of v_f/v_t varied approximately with feed location as follows: (1) 6B DISK, Radial Feed: 0.2; (2) 6B DISK, Above Impeller Feed: 0.05; (3) HE-3, Radial Feed: 0.04 and (4) He-3, Above Impeller Feed: 0.03.

INTRODUCTION

Many experimental studies have shown ([1], [2], [3] and see [4] for an extensive literature review) that the yield of a primary reaction for fast competitive or competitive/consecutive (C/C) reactions is enhanced by rapid blending of feed with the reaction batch. Rapid feed blending can be achieved by (a) increasing the agitator speed at a given feedpipe location or by (b) moving the feedpipe into the most intensively turbulent regions of the vessel, which are located nearest the impeller blade tips. Unfortunately, the rapid feed blending achieved by feeding near the impeller blade tips is not achieved without the possible detrimental effects of backmixing into the feedpipe. The same high levels of turbulence which promote rapid blending will also promote backmixing into the feedpipe. Backmixing of reactor contents into the feedpipe will result in reaction within the feedpipe where blending is much slower than the blending achieved within the vessel and where non-stoichiometric conditions may promote competing side reactions.

For fast C/C reactions in agitated vessels backmixing into the feedpipe has been shown by Bourne et al [1] to lower yields of desired product and increase the yield of undesired product. Bourne and Rohani [2] have speculated that feedpipe backmixing contributed to lowered yields in the study of Paul and Treybal [3].

There are no quantitative design methods for predicting the onset of or degree of backmixing into feedpipes; although Fasano and Penney [4] have given preliminary recommended feedpipe velocities to eliminate feedpipe backmixing. But those recommendations are based on the knowledge of turbulent intensities in agitated vessels and not on experimental evidence of feedpipe backmixing.

The objective of this study was to determine operating conditions for which feedpipe backmixing can be eliminated in agitated vessels.

REVIEW OF LITERATURE

Bourne, Kozicki, Moergeli and Rys [1] were the first investigators to demonstrate that feedpipe backmixing can have a detrimental effect on the yield of fast C/C reactions in agitated vessels under fedbatch and continuous operation. In the absence of feedpipe backmixing, an increase in agitator speed decreases the yield of the C/C product and increases the yield of primary product. The effect of feedpipe backmixing was shown to increase the yield of a C/C product as agitator speed was increased beyond the speed where feedpipe backmixing occurred. The effect of feedpipe backmixing was dramatic with the yield of the C/C product being increased 3 to 10 fold by the effect of feedpipe backmixing, at the highest speeds used. Bourne [1] et al also demonstrated that feedpipe backmixing resulted at high agitator speeds by observing color formation within the feedpipe when the color resulted from the formation of a colored C/C product.

They also determined conditions where feedpipe backmixing did not occur. Feedpipe backmixing was eliminated by increasing the feedpipe velocity. For the feed location used (which was D/4 below the 6B DISK impeller and at a radial location even with the inboard tip of the blades) at $v_f/v_t = 0.333$ feedpipe backmixing was eliminated as evidenced by no color formation in the glass feedpipe. Bourne et al [1] also conducted tests with v_f/v_t varying from 0.004 to 0.15 where feedpipe backmixing occurred as evidenced by color formation in the glass feedpipe or from an increase of C/C product with an increase in agitator speed.

From the results of Bourne et al [1], for the 6B DISK impeller and the feed location used, it appears that v_f/v_t must be above 0.3 in order to eliminate the effect of feedpipe backmixing.

Bourne and Rohani [2] have analyzed the data of Paul and Treybal [3] and they have commented about the possibility of feedpipe backmixing affecting the results, "Using feed position 'a' (0.00635 m (0.25 inch) from the 6B DISK in the discharge stream) there was a risk of forcing the high velocity jet of tyrosine solution leaving the turbine into the 3.2 mm (0.125 inch) iodine feed pipe. The velocity of the iodine solution was 0.05 m/s (0.164 ft/s), whereas depending upon the turbine size and speed the discharge had a radial velocity component of order 0.2 to 1.2 m/s (0.66 to 3.94 ft/s) (turbine tip speeds 0.5 to 3 m/s). Such backmixing has been observed in an analogous case [1]. Had it occurred here, the yield of R would have been reduced."

As Bourne and Rohani [2] indicated, Paul used a velocity ratio (v_f/v_t) range of 0.017 to 0.1. And based on the experimental results of Bourne et al [1], and on the recommendations of Fasano and Penney ($v_f/v_t > 0.5$), it is very likely that most of the experimental data of Paul and Treybal [3] were affected by backmixing into the feedpipe. However, the quantitative effect of the feedpipe backmixing on their data cannot be determined. The effect of feedpipe backmixing on Paul's data could have been less than the effect on Bourne's data because Paul's feedpipe was relatively smaller than Bourne's (T/d = 40 for Paul's work and T/d = 25 for Bourne's work). Thus, based on geometric considerations Paul's feedpipe volume ratioed to the vessel volume would have been about 1/4 the relative feedpipe volume used by Bourne.

EXPERIMENTAL EQUIPMENT

A sketch of the vessel and the experimental apparatus is included as Figure 1. A sketch of the feedpipe is included as Figure 2. The vessel was constructed of a Plexiglass cylinder of 0.254 m (10 inch) outside diameter and 0.00635 m (0.25 inch) wall thickness.

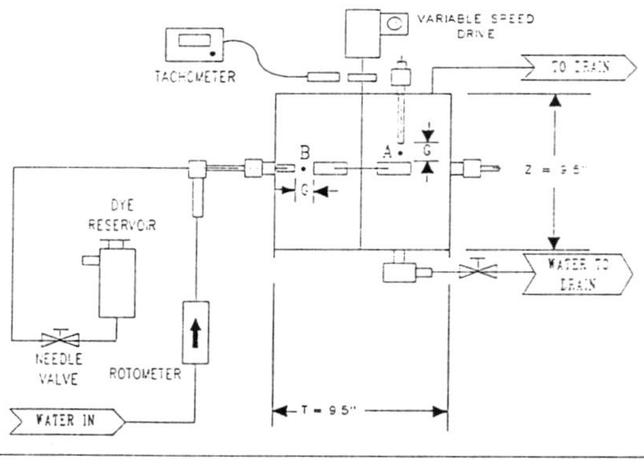

Figure 1. Experimental Apparatus for Feedpipe Backmixing Investigations.

The tank was operated liquid full. The top plate was 0.0254 m (1 inch) thick plexiglass. The tank could be drained. The water which entered the tank through the feedpipe was routed to the drain through an overflow pipe attached to the top plate of the vessel. The agitator speed was measured with an Airpax (Model 77220-1-201) digital tachometer. A Schutte-Koerting rotameter was used to measure the water flowrate. A 0.0635 m (2.5 inch) diameter by 0.3048 m (12 inch) long stainless steel pressurized reservoir was used to hold the dye solution. The dye solution was metered into the center of the feedpipe by adjusting a needle valve in the line between the reservoir and the feedpipe. The dimensions of the 6B DISK and the HE-3 impeller(s) are given in Table 1.

The feedpipe (Figure 2) was 0.0127 m (0.5 inch) outside and 0.009525 m (0.375 inch) inside diameter glass tubing. The dye solution entered the feedpipe on its centerline 0.00635 m (0.25 inch) from its end through a plastic tube of 0.003175 m (0.125 inch) outside diameter and 0.002159 m (0.085 inch) inside diameter. The glass tubing was inserted through the vessel wall through a Swagelok[R] tube fitting. The feedpipe was either positioned pointing towards the discharge of the impeller (at the impeller midplane, position B) or above the impeller at a radial location of 0.0381 m (1.5 inch) from the tank vertical centerline (position A).

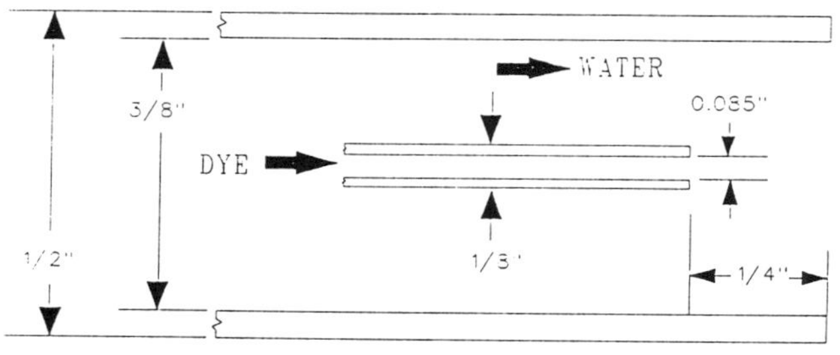

Figure 2. Details of the end of the Feedpipe.

Table 1. DIMENSIONS OF IMPELLERS

IMPELLER STYLE	D (m)	L (m)	W (m)	D_d (m)	d_s (m)	L/D	W/D	D_d/D
6B DISK	0.0889	0.0178	0.00223	0.0589	0.0127	0.200	0.251	0.663
	0.127	0.0254	0.03175	0.0843	0.0127	0.200	0.250	0.664
HE-3	0.127	Standard			0.0127			

EXPERIMENTAL PROCEDURE

Tap water was used as the vessel fluid and as the feedpipe fluid. At the start of an experiment the vessel was filled with clean water, the feedpipe water flow rate was set and the agitator was started and set at a low speed. The dye solution was gradually increased by opening the needle valve until the dye stream could be clearly seen through the wall of the glass tube. If no dye migrated backward against the flow through the feedpipe, then it was assumed no feedpipe backmixing was occuring. Then the agitator speed was gradually increased until dye finally migrated back along the feedpipe counter to the feedpipe flow. By definition this was the condition at which feedpipe backmixing occurred.

The identification of feedpipe backmixing was slightly affected by the velocity of the dye solution exiting the dye injection tube and by the initial dilution of the food dye. The dyes used were full strength Red and Leaf Green food dyes manufactured by Chefmaster Products. When diluted 3:1 with tap water the Leaf Green dye (consisting of Yellow #5 and Blue #1) gave the best visual results. At this dilution the dye injection velocity could be held below the feedpipe velocity, the viscosity was sufficiently low so that the dye readily dispersed in the water and the color was sufficiently intense so that backmixing could be visually detected. At too high injection velocities, the dye jet penetrated from the tube into the vessel and the jet effect could partially mitigate the backmixing effects due to agitation. With full strength dye, the injection velocity could be held low but the higher viscosity of the dye retarded the thread of dye dispersing and following the turbulence and thereby migrating against the feedpipe flow.

RESULTS

Table 2 presents a summary of selected experimental conditions for both the 6B DISK impeller and the HE-3 impeller. The data taken were agitator speed and water flow through the rotameter. These data were reduced and the following parameters were calculated: N_{Rei}, N_{Ref}, v_t, v_f, G/D and D/T. For an impeller and a feedpipe location a linear correlation of v_f versus v_t was observed. Figures 3 thru 7 present these data for 16 combinations of impeller and feedpipe location. The Figure numbers for the combinations are given in Table 2. The data were regressed to a linear fit. Some scatter of the data would be expected as the feedpipe Reynolds number (N_{Ref}) varied from a low of 100 to a high of 1300 so the feedpipe flow regime was laminar or transitional. The velocity variation over the tube would be expected to affect the feedpipe backmixing. The dye was injected where the feedpipe local velocity is highest. This could

Feed Pipe Backmixing
3.5" 6BD, Feed Location B

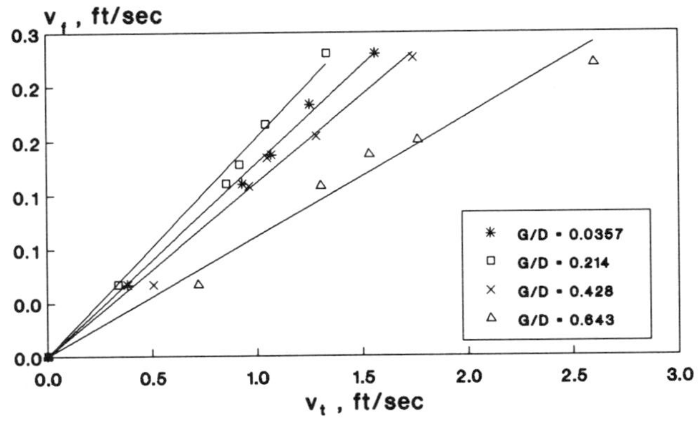

Figure 3

Table 2. SUMMARY OF EXPERIMENTAL CONDITIONS AND RESULTS

IMPELLER STYLE	FEEDPIPE LOCATION G (m)	POS.	D (m)	G/D	D/T	RANGE N_{Rei}	RANGE N_{Ref}	v_f/v_t	FIGURE NUMBER
6B DISK	0.003175	B	0.0889	0.036	0.37	2-20E3	1.0-10E2	0.1796	3
"	0.0127	B	"	0.143	"	2-30E3	1.1-13E2	0.1968	3
"	0.0254	B	"	0.286	"	3-32E3	1.2-13E2	0.1884	3
"	0.0508	B	"	0.571	"	5-36E3	0.7-10E2	0.1280	3
"	0.003175	B	0.127	0.025	0.53	2-24E3	0.8-10E2	0.1702	4
"	0.0127	B	"	0.100	"	3-24E3	1.1-10E2	0.1782	4
"	0.0254	B	"	0.200	"	2-18E3	0.6- 8E2	0.1866	4
"	0.0508	B	"	0.400	"	3-28E3	0.7- 8E2	0.1344	4
HE-3	0.003175	B	0.127	0.025	0.53	4-75E3	0.4- 7E2	0.0552	5
"	0.0127	B	"	0.100	"	1-12E4	0.5- 8E2	0.0263	5
"	0.0254	B	"	0.200	"	1-12E4	0.5- 7E2	0.0265	5
"	0.0508	B	"	0.400	"	1-11E4	0.5- 6E2	0.0209	5
6B DISK	0.0254	A	"	0.200	"	6-84E3	0.9-10E2	0.0478	6
"	0.028575	A	"	0.225	"	6-84E3	0.9-10E2	0.0480	6
"	0.041275	A	0.0889	0.464	0.37	3-12E4	1.7- 7E2	0.0249	7
HE-3	0.0381	A	0.127	0.300	0.53	2-11E4	1.0- 7E2	0.0267	8

Feed Pipe Backmixing
5" 6BD, Feed Location B

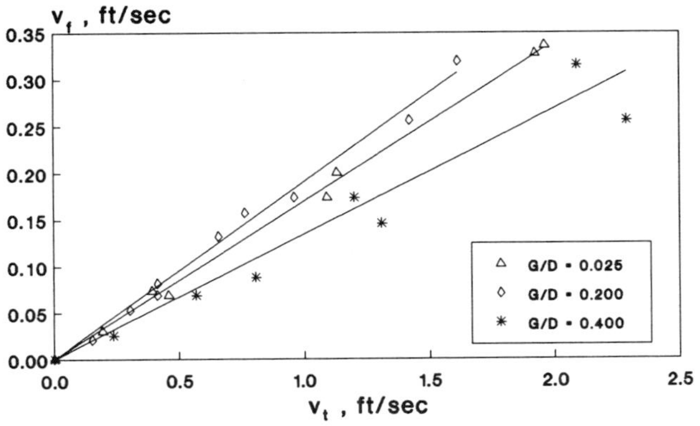

Figure 4

Feed Pipe Backmixing
3.5" 6BD, Feed Location A

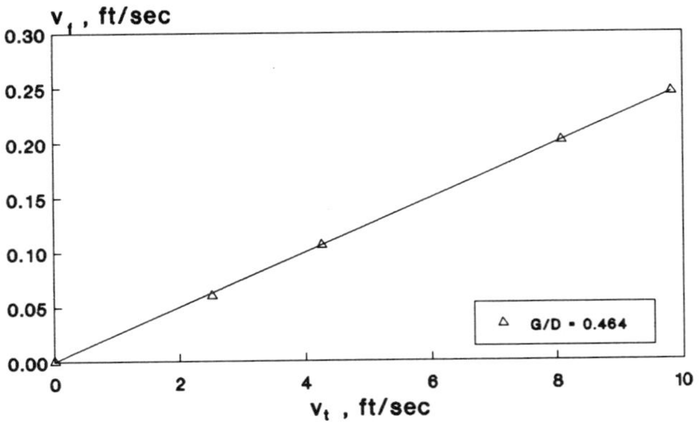

Figure 7

Feed Pipe Backmixing
5" HE-3, Feed Location B

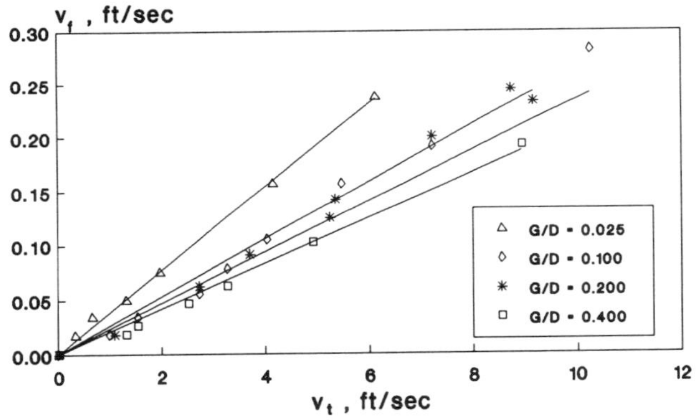

Figure 5

Feed Pipe Backmixing
3.5" HE-3, Feed Location A

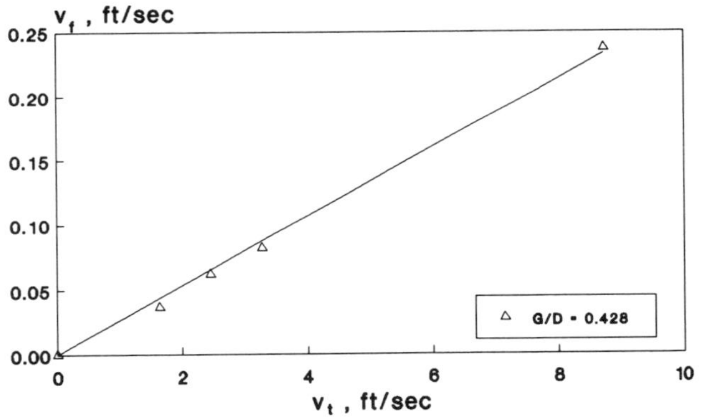

Figure 8

Feed Pipe Backmixing
5" 6BD, Feed Location A

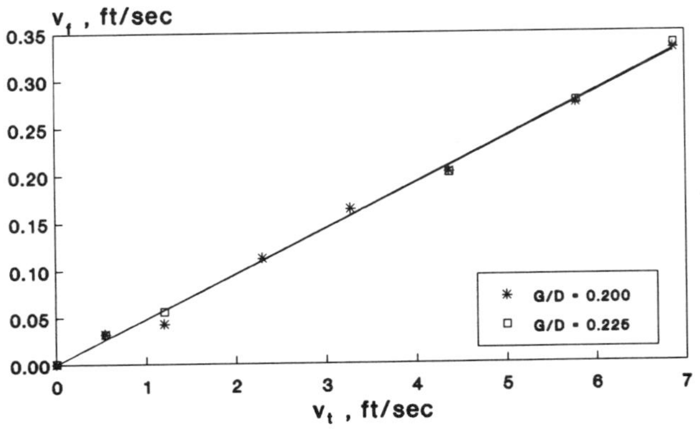

Figure 6

perhaps increase the impeller speed needed to produce backmixing over the speed required for the flatter velocity profile for fully turbulent feedpipe flow. However, the normal mode of backmixing was for dye to either leave or almost leave the end of the feedpipe and was then swept by the vessel turbulence back into the feedpipe near the feedpipe wall where the local velocity is lowest. Experiments need to be conducted under conditions where the feedpipe flow is fully turbulent. This was not possible for this investigation because at higher flow rates the vessel turbulence was so high that it was difficult to follow the feedpipe backmixing with the naked eye.

From a design standpoint what one needs to know is the feedpipe velocity needed to avoid backmixing. The ratio of v_f/v_t for the various combinations of impeller and feedpipe location is given in Table 2. The ratio varies from a high of about 0.2 for 6B DISK, G/D = 0.2, feed location B, to a low of 0.025 for 6B DISK, G/D = 0.46, feed location A. For the HE-3 impeller Figures 9 and 10 present v_f/v_t versus G/D for the 0.0889 m (3.5 inch) and 0.127 m (5 inch) 6B DISK and the 0.127 m (5 inch) HE-3, respectively.

Feed Pipe Backmixing
6BD Impeller, Feed Location B

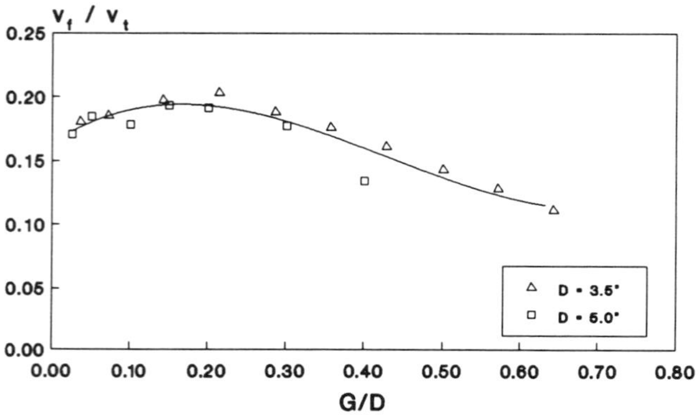

Figure 9

Feed Pipe Backmixing
5" HE-3 Impeller, Feed Location B

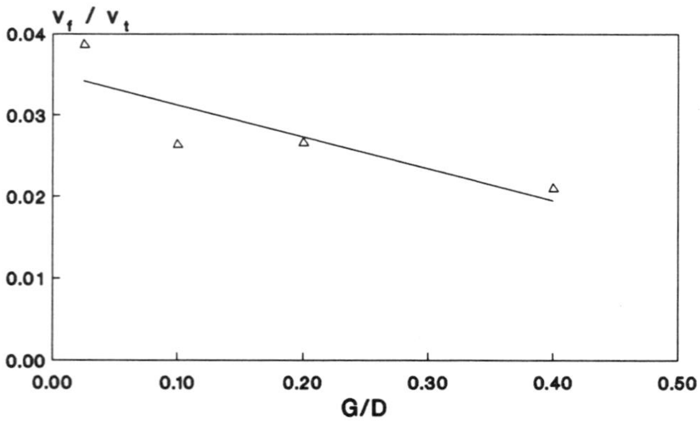

Figure 10

ANALYSIS OF RESULTS

The range of the experimental values are presented in Table 2. Additionally, D/d = 25.3. For geometrically similar systems v_f/v_t should be independent of scale. For systems which are not geometrically similar the values of v_f/v_t must be selected based on judgement and the information presented in Table 2.

The allowable v_f of Bourne et al [1] are higher than the findings of this study. They determined from the feedpipe color experiments that feedpipe backmixing did not occur at v_f/v_t = 0.33 for D/d = 150, D/T = 0.3 and G/D = .25, feed location A. This is higher than the value of 0.05 that was obtained here for G/D = 0.2. Bourne et al [1] found that backmixing occurred during reaction runs down to a v_f/v_t of about 0.143. Thus, this study found v_f allowable about 1/3 that of Bourne's study.

CONCLUSIONS AND RECOMMENDATIONS

Feedpipe backmixing does occur above a critical agitator speed for any feedpipe location as evidenced by the visual determination of injected color migrating counter to the flow in the feedpipe.

Feedpipe backmixing was determined for two diameters of 6B DISK impellers and one HE-3 with feedpipe locations at various distances radially out (Location B) from the impeller and for locations above the impellers (Location A).

The visual method of determining feedpipe backmixing with injected color is most successful at low agitator speeds where the level of turbulence in the vessel is relatively low. For high agitator speeds the turbulence level in the vessel is so high that the dye is dispersed so rapidly that the naked eye cannot follow dye penetrating back into the feedpipe. For higher v_f and v_t some method such as conductivity will need to be used. Backmixing data need to be taken for turbulent conditions in the feedpipe.

The data obtained here indicate that, for a given geometry, v_f/v_t, at the critical agitator speed for backmixing to occur, remains relatively constant as agitator speed is changed.

For design purposes when determining feedpipe velocity for systems geometrically similar to those studied here, the authors recommend that the design velocity be twice the predicted feedpipe velocity. This will allow for any increase in v_f/v_t with scaleup.

NOTATION

Roman Letters

d = Feedpipe Diameter
d_s = Shaft Diameter
D = Impeller Diameter
D_d = Impeller Disk Diameter
G = Gap between Feedpipe and Impeller
L = Impeller Blade Height
N = Impeller Rotational Speed
N_{Rei} = Impeller Reynolds Number = $ND^2\rho/\mu$
N_{Ref} = Feedpipe Flow Reynolds Number = $v_f d\rho/\mu$
T = Vessel Diameter
v_t = Impeller Tip Speed
v_f = Feed Flow Velocity
W = Impeller Blade Width
Z = Vessel Height

Greek Letters

ρ = Fluid Density
μ = Fluid Viscosity

LITERATURE CITED

1. **Bourne, J.R., Kozicki, F., Moergeli, U. and Rys, P.,** "Mixing and Fast Chemical Reaction - III," Chem. Eng. Sci., Vol. 36, No. 10, pp. 1655-1663 (1981).

2. **Bourne, J.R. and Rohani, S.,** "Micro-Mixing and the Selective Iodination of l-Tyrosine," Chem. Eng. Res. Dev., Vol. 61, pp. 297-302 (Sept. 1983).

3. **Paul, E.L. and Treybal, R.E.,** "Mixing and Product Distribution for a Liquid-Phase, Second-Order, Competitive-Consecutive Reaction," AIChE J., 17, No. 3, pp. 718-724 (1971).

4. **Fasano, J.B. and Penney, W.R.,** "Cut Reaction By-Products by Proper Feed Blending," Chem. Eng. Prog., 89, pp. 46-52 (Dec. 1991)

Application of Electrical Impedence Tomography to Mixing in Stirred Vessels

R. A. Williams, R. Mann, F. J. Dickin, O. M. Ilyas, and P. Ying

Department of Chemical Engineering, UMIST, Manchester, U.K.

R. B. Edwards

Unilever Research, Port Sunlight Laboratories, Merseyside, U.K.

A. Rushton

British Nuclear Fuels PLC, Preston, U.K.

Two-dimensional non-invasive scanning or tomography has become an important part of modern medicine and has as yet unrealized potential for application to chemical process units. A series of electrodes (16-32) mounted on the periphery of a stirred vessel provide a non-invasive method of monitoring the resistivity of fluids inside. A 5 mA current injected through a pair of neighboring electrodes is conducted through the fluid and the resulting voltages are measured at all other neighbor pairs. This procedure is repeated for all electrode pair combinations. The measurements exploit every possible combination of electrode pairs, taking only 79 ms to give a total of 104 views of a section of the mixing vessel. A program running on a personal computer is able to map resistivity and reconstruct images at a rate of one frame per second, producing a detailed map of the mixing process as it occurs. A series of results is presented for a single impeller configuration, mapping out the batch mixing of a slug of brine tracer. Simultaneous capture of video images of the tracer mixing provides a visual confirmation of the tomographic reconstructions. Results have been modelled using the networks-of-zones mixing model. The large amount of data generated which shows the detail of the concentration changes in space and time as mixing proceeds. Electrical impedance tomography (EIT) provides a most appropriate experimental basis for correlating the detailed behavior predicted by computational fluid dynamics (CFD).

INTRODUCTION

Tomography can be simply described as a methodology of obtaining 2-D information in a non-invasive manner. This sort of technique was first developed in the 1930's to diagnose lung disorders. Since then tomography has become an acceptable technique for medical diagnosis purposes (Bates et al, 1983). Nevertheless this type of technology has only recently (1980's) become developed to an acceptable level to allow the investigation of complex and dynamic chemical processes for which it has hitherto been difficult to obtain reliable data. Examination of the literature (Ilyas, 1991) reveals tomographic techniques have been used for numerous applications (fluidized bed studies, multiphase flow imaging, bioreactors, temperature mapping), employing a wide range of tomographic sensing techniques (e.g. nucleonic (X-ray, γ-rays), optical, ultrasonic and electrical methods). Many of these methods have been developed to their full potential in the medical field but some of the methods are less suited for process applications such as optical sensors (redundant in opaque systems), and radiation based sensors (slow, expensive and have safety constraints). The work described here is part of on-going research at UMIST based on the use of electrical sensors. Earlier work developed a capacitance tomographic instrument suitable for non-conducting systems (Salkeld et al, 1991). The current impetus is to develop an electrical impedance system to be used for conducting systems. Tomographic systems typically comprise a multiple assembly of external radial arrays of sensing points. Whenever such arrays are reasonably large, by appropriate reconstruction techniques, quite fine detail across any section can be resolved. The level of detail resolved obviously justifies modelling approaches which are capable of providing a matching degree of density of information. In this respect, the network-of-backmixed zones concept provides for a level of spatial discrimination which is commensurate with the potential of electrical impedance tomography (Mann and El-Hamouz, 1991).

In this paper a demonstration of an axi-symmetric tracer test monitored by electrical impedance tomography (EIT) will be presented, which can be simulated by a 2-D networks-of-zones model and simultaneously confirmed by direct visualisation(using selected video frames). Such verification schemes are necessary to validate and check the EIT equipment, methodology and algorithms. A simple asymmetric single point injection result will then be presented, which shows that it is just as straight forward to use EIT to collect information in 3-D using the same equipment. Interpretation of the non-symmetric case, which is most relevant to conventional industrial practise, demands the development of fully 3-D modelling approaches. These are currently under development at UMIST.

ELECTRICAL IMPEDANCE TOMOGRAPHY (EIT)

The objective of EIT is to obtain a 2-D map of the resistivity (inverse of electrical conductivity) distribution in the imaging plane. The results presented in this paper were derived using an instrument adapted from a medical impedance instrument (Barber and Brown, 1984).

Figure 1 shows the experimental set up for an approximately 30 dm^3 perspex vessel fitted with a *standard* Rushton turbine and four equi-spaced baffles (Holland, 1962). This test vessel was fitted with four concentric sets of electrodes. Two upper sets were spaced above the impeller and one set below it, with a further set in the plane of the impeller, as depicted in Figure 1.

In essence, the tomographic system for such a typical process vessel then comprises three basic components, as indicated in Figure 2, which are:

(a) Electrical sensors, which in this case consist of 16 electrodes (in the form of stainless steel bolts inserted into the walls) that are attached to the periphery of the process vessel in a non-intrusive fashion.

(b) A data collection unit, consisting of current sources, voltmeters, and a control circuit to control the interaction of current and voltage.

(c) A computer, which is used to collect the data, process and display the image.

Data is collected by injecting a 5 mA peak a.c. current via a neighbouring pair of electrodes and measuring the voltage at the other adjacent electrode pairs. Figure 2 illustrates the methodology: current is injected via electrodes 1-2, the voltage is measured at electrodes 3-4, 4-5, .. 15-16. then the current driving pair is then changed to electrodes 2-3, and similar voltage measurements are obtained. This protocol is repeated for all possible electrode combinations (1-2, 2-3 .. 16-1), this is done in 79 ms. This provides 104 unique voltage measurements from which a resistivity image can be deduced using an appropriate reconstruction algorithm. However using this data acquisition protocol results in a non-uniform current density, i.e. good sensitivity near the periphery of the vessel and poorer sensitivity in the centre.

The reconstruction algorithm used in this work is simple to implement, and capable of reconstructing images quickly (1 frame per second on a 186 PC); it is referred to as back-projection between equipotential lines. Figure 3 illustrates the method pictorially, assuming a square geometry vessel or pipe, which has three sensors and detectors in the X and Y directions. The shaded area in the top 3 diagrams have one or more shaded areas which represent, for instance, a gas bubble. Looking in the X and Y direction the object can be located from the overlap of the two measurement signals (X and Y) in cases 1 and 2. However, in case 3 the procedure gives an incorrect result, this implies that more than two views are needed (104 in this case). The back projection in this case is performed along the equipotential lines; Figure 1 shows a shaded region between equipotential lines terminating at electrodes 6-7, which is one of the 104 views used . An inherent feature of this algorithm is the noise generated due to the overlapping of the views, this is usually remedied by using appropriate filters. The *resolution* of the image is usually insufficient to allow small particles or regions to be observed (Dickin et al, 1993), but the average resistivity within a pixel can be revealed, for example, the proportion of two different components within a process mixture.

MIXING EXPERIMENTS

The mixing of miscible aqueous liquids was investigated by introducing the secondary liquid either as a pulse from a symmetrical set of 4 syringes mounted above the liquid (as shown in Figure 1(b)) or by continuous flow from a burette.

The vessel was filled with a primary liquid (0.1 wt% NaCl solution, 1.9 ms cm^{-1} at 18°C) to which an aliquot of the more concentrated secondary solution (1 wt% NaCl, 18.2 ms cm^{-1} at 18°C) was added. The volume(s) of the aliquots varied from 8 to 22 ml according to the particular experiment. Two types of secondary liquid addition were employed:

(i) axisymmetric addition in which the secondary liquid was introduced from 4 syringes mounted symmetrically around the stirrer shaft above the liquid level in the vessel. The tops of the 4 syringes projected just below the liquid surface. This represents an arrangement for an approximately symmetric injection of the more concentrated brine.

(ii) asymmetric addition in which the secondary liquid was added from 1 syringe only (as a pulse) or alternatively in a semi-batch fashion continuously from a single burette positioned midway across the radius of the tank.

Experiments were performed to examine the influence of agitation rate (in terms of the rpm of the stirrer) and the symmetry of the addition of the secondary liquid. Parallel experiments were performed using injections of a Nigrosine dye-tracer, in place of the secondary liquid, and observing the resultant dispersion of the dye using a video camera. This provides complementary two-dimensional (vertical) information based on *external* observations of the tank for comparison with the two dimensional (horizontal) information derived from the *internal* observations facilitated by electrical impedance imaging. Theoretical simulations of the same experimental were also performed using a network simulation model, described below.

EXPERIMENTAL RESULTS - AXISYMMETRIC PULSE INJECTION

Figure 5 presents a set of tomographic images for the injection of 4 ml of the more concentrated brine at the four vertically spaced planes fitted with radial arrays of impedance sensors.

The stirrer speed is 100 rpm.

At $t=0$, just before injection, each plane shows an approximately uniform impedance as deduced from the back projection algorithm. There is some 'noise', in that each plane is not perfectly uniform as would be expected for a uniformly mixed fluid (in this case 0.1% brine). The 'noise' is greatest close to the impeller.

After approximately instantaneous injection at $t=0$, after 1 s the tomographic images show clearly the changes in local resistivity arising from the injection of the more concentrated 1% brine. According to the set of tomographic images at 1 s, the more concentrated brine is most detectable close to the impeller. There is some patchiness in the local deduced resistivity, probably indicative of the 'four' slugs of 1% brine entering the fluid from the four separate nozzles. This suggests that the tomographic technique is capable of good resolution. The captured video image at 1 s, using nigrozine dye to make the pulse of brine visible, represents a 'forward' view of the cloud of tracer, which suffers from a certain amount of self-obscuration. Thus, although the visual image cannot resolve to the same detail, it provides a good verification of the general state of mixing indicated by the tomograms. After 2 s, the extremes of concentration of the brine have attenuated according to the tomographic images and there is now a detectable increase at the lowest plane below the impeller. Some asymmetry of the mixing is suggested from the tomography, although this cannot be precisely verified on the visualisation. The visualised image at 2 s shows a low dye concentration at the vessel surface and this is also borne out by the tomography. The tomography result also suggests some swirl effect, with brine precessing preferentially along the wall. Again, this effect cannot be seen in the self-obscuring forward view of the dye visualisation.

After 3 s, the tomographic picture is becoming less distinct as homogenisation is beginning to be approached. This approach to complete mixing is certainly indicated in the visualisation case.

THEORETICAL ANALYSIS USING NETWORKS-OF-BACKMIXED-ZONES

In the work reported here, experiments have been conducted using an impeller at a clearance above the base of $C=H/3$ for a standard geometry tank with $H=T$. It is thus necessary to modify the former 'standard' single impeller. However, unlike previous work using the model at one-third clearance (Mann and El-Hamouz, 1991), instead of having an N'xN' network below the impeller and an N'x2N' arrangement above the impeller (where all the cells are of equal volume), this development has retained the simpler (N'xN') + (N'xN') algorithm, but reduced the volume of the cells below the impeller by 2/3 and correspondingly increased the volume of the cells above the impeller by the factor 4/3.

Figure 6 shows the schematic arrangement of the zones. Note the ones above the impeller have been *stretched* to accommodate the adjustment in zone volumes. To indicate the general N'xN' network structure, the outer loops of zones in this figure are shown partially complete. The impeller is located between $j=N'$ and $j+N'+1$. The focus of the lower circulation loops is between $i=N'/2$ and $i=N'/2+1$ and $j=N'/2$ and $j=N'/2+1$. The upper loop focus is then between $i=N'/2$ and $i=N'/2+1$ and $j=(3/2)N'$ and $j=(3/2)N'+1$. For the present case, N' has been chosen to be 20, which gives an overall total of 800 zones. This number represents a good compromise between having a reasonably fine spatial discrimination, without excessive computing time.

Each cell undergoes an equal and opposite exchange flow with adjacent cells in the lateral inner and outer flow loops which imitates the effect of turbulent exchange between parallel flowing streams of fluid which circulate by the action of the impeller. A turbulent exchange flow parameter of $\beta=0.3$ provides accurate simulations of the experimental dispersion of inert and reactive tracers. Figure 7 shows the results of simulations using the above networks of size $N'=20$. The model predicts the evolution in space and time of the concentration in each of the 800 zones. The results in Figure 7, for half-second internals, have been produced by constructing the concentration contours and shading in between them in accordance with the scales indicated on the Figure.

The predictions in Figure 7 show a very good correspondence with the visualisations in Figure 5, if the limit of visual detection of the nigrozine dye is taken to be at a relative concentration of 1.5. Figure 7 comprises half-section image reconstructions of the axisymmetric concentration fields, which were previously produced using a laborious line shading method (Mann and El-Hamouz, 1991; Wang and Mann, 1992). The predicted convection and turbulent dispersion of the inert slug of dye is exactly in line with these networks-of-zones simulations. The assumption of perfect symmetry means that the model cannot reproduce the patchiness indicated by the tomographic images, which in any case is bound to be obscured by the forward view method used in the direct visualisations.

Nevertheless, it can be concluded that the network-of-zones model provides for a scale of scruting of the tomographic images which matches their capacity for providing very detailed information on the shape and structure of the evolving concentration fields in a batch tracer injection test.

EXPERIMENTAL RESULTS - ASYMMETRIC PULSE INJECTION

Figure 8 shows a photograph of a time sequence of images for measurements performed at one cross-section (17 cm above the base of the tank), for asymmetric addition of the secondary liquid for three different impeller speeds. The major features of interest can be seen from these images (originally in colour but reproduced here in mono-tone).

Considering the sequence for 50 rpm, the first images (time = -1 s) is a 'reference' image obtained with the tank containing the primary liquid and whilst agitating the liquid. The second image (time = 0) corresponds with the time when the secondary liquid has been added, hence large changes in conductivity are perceptible from the image (the region of highest conductivity is the white core surrounded by a darker area). With continued mixing this high conductivity zone is dispersed until after 10 s no further statistical changes in the image can be deduced. Nevertheless having achieved a homogeneous mixture some artifacts associated with the conducting central impeller and wall conduction are seen on the final image.

For tests performed at higher mixing speeds, some 'noise' is detected within the reference image (obtained with the primary liquid and stirred at the appropriate rate). On addition of the aliquot of concentrated solution, it is dispersed rather more effectively, and the corresponding times to achieve mixing to a homogeneous state were 14 s and 8 s for impeller speeds of 100 rpm and 200 rpm, respectively.

Hence, the tomographic method offers a convenient way to assess the mixing times required to achieve a uniform distribution of the secondary liquid. However, much more importantly, Figure 8 provides fully detailed information on the non-symmetrical mixing case. This is typical of much industrial practice, where a second fluid is added from a single dip-pipe. The mixing that accompanies this single point addition is inherently non-symmetrical, even if the fluid flow and mixing were themselves intrinsically perfectly symmetrical (which there are almost certainly not, although the degree of tangential swirl in the third dimension is usually much restricted by the use of baffles). Figure 8 therefore confirms that the experimental acquisition of fully 3-D concentration field information is just as straight forward as for the experimentally artificial 2-D symmetry case examined in Figures 5 and 7. It can therefore be concluded that Figure 8 provides a powerful incentive to develop and implement fully 3-D models of the flow and mixing. Moreover it is important to emphasise that the detail precisely commensurate with the detail that is capable of being provided from computational fluid dynamic (CFD) models of which a 3-D network-of-zones would represent a somewhat simplified representation. Full solutions to the time averaged Navier-Stokes equations, typically achieved through commercially available software packages, such as the PHOENICS example, can now be pursued with the recognition that the wealth of detail that can be produced from the fluid dynamics and mixing can in turn be subject to experimental verification by electrical impedance tomography (EIT).

REFERENCES

Barber, D.C. and Brown, B.H. (1984), J.Phys. E: Sci. Instrum., **17**.
Bates, R.H.T., Garden, K.L., Peters, T.M. (1983), IEEE Trans., **71**.
Dickin, F.J., Williams, R.A. and Beck, M.S. (1993), Chem. Eng. Sci. (in press).
Holland, A.F. (1962), Chem. Eng., **17**.
Ilyas, O.M. (1991), MSc Thesis, University of Manchester Institute of Science and Technology.
Mann, R. and El-Hamouz, A.M. (1991), Proc. 7th Europ. Conf. on Mixing, **1**.
Salkeld, J.A., Hunt, A., Thorn, R., Dickin, F.J., Williams, R.A., Conway, W.F. and Beck, M.S. (1991) in "Advances in Measurement & Control of Colloidal Processes", Butterworth-Heinemann (London).
Wang, Y-D and Mann, R. (1992), Trans IChemE, **70(A)**, 282.

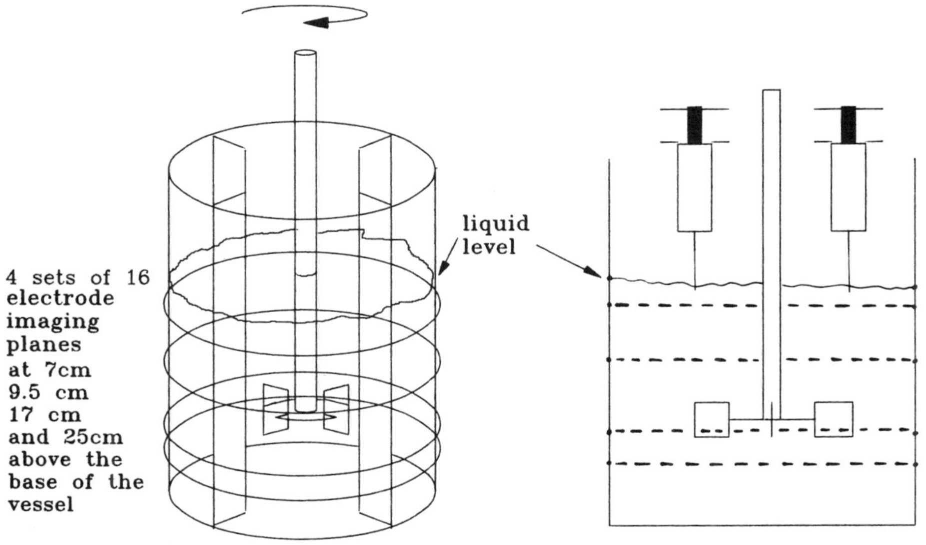

FIGURE 1 Stirred Vessel and Imaging Planes

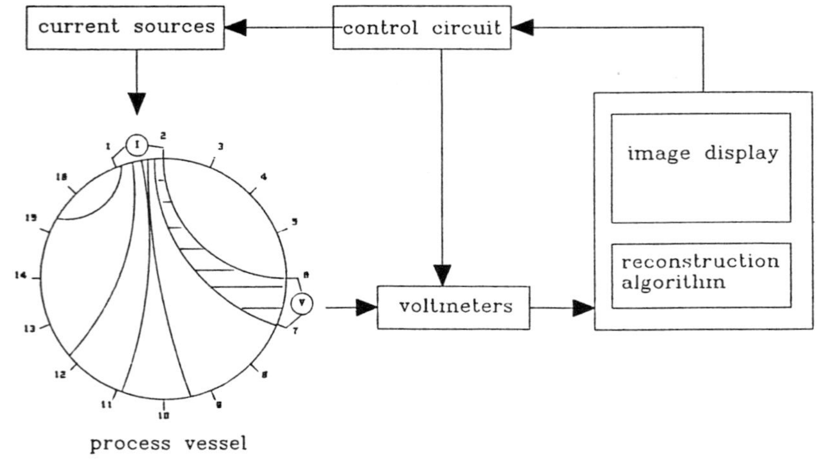

FIGURE 2 Schematic Diagram of Impedance Tomography System

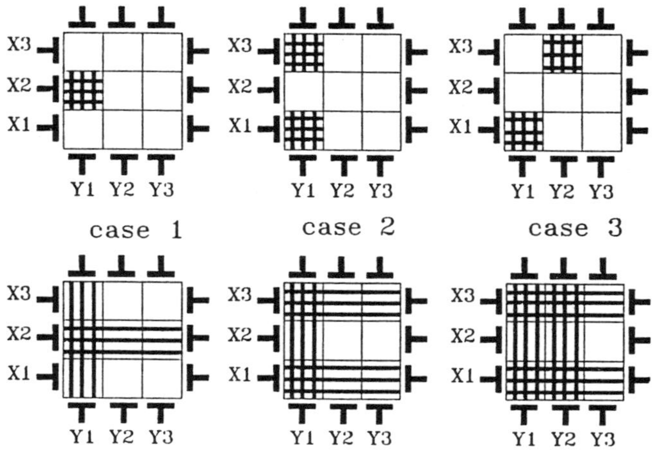

FIGURE 3 Principle of Object Recognition using Back-Projection

FIGURE 4 View of Stirred Vessel Showing Impedance Measuring Arrays

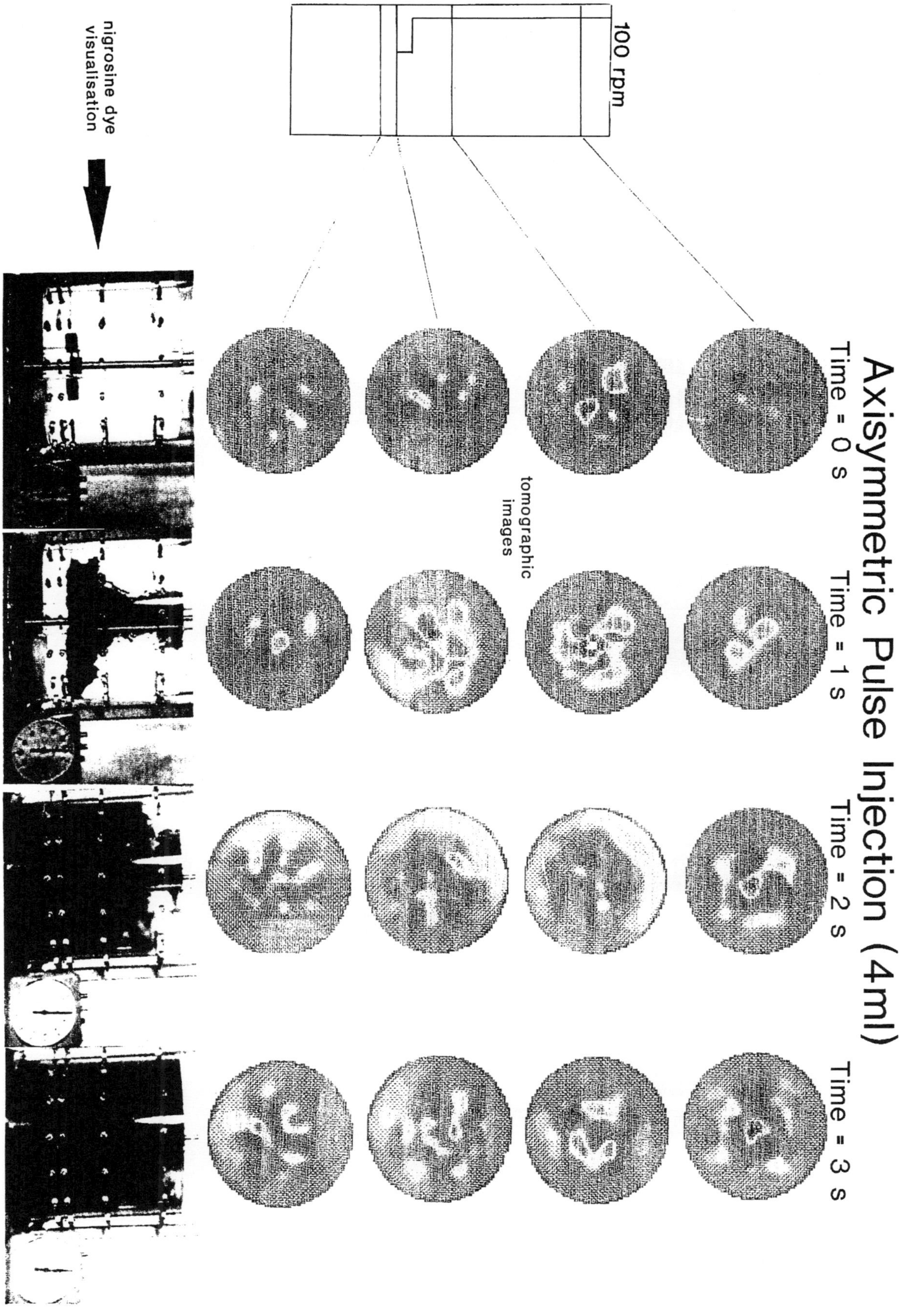

FIGURE 5 EIT Images and Direct Visualisation for Symmetric Injection

FIGURE 6

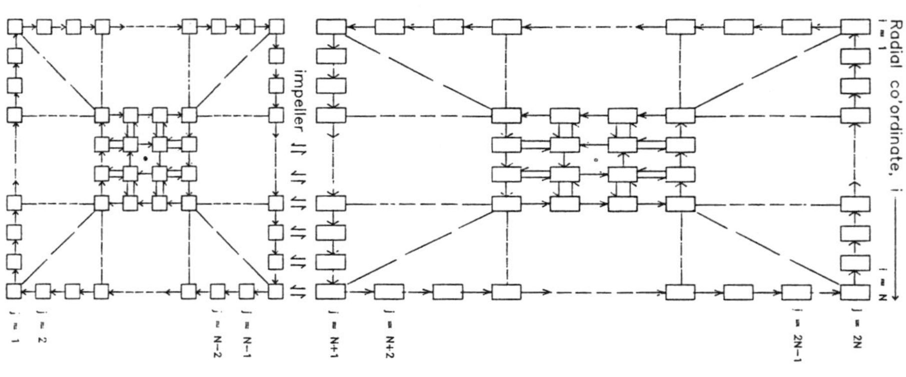

Network-of-Zones Model

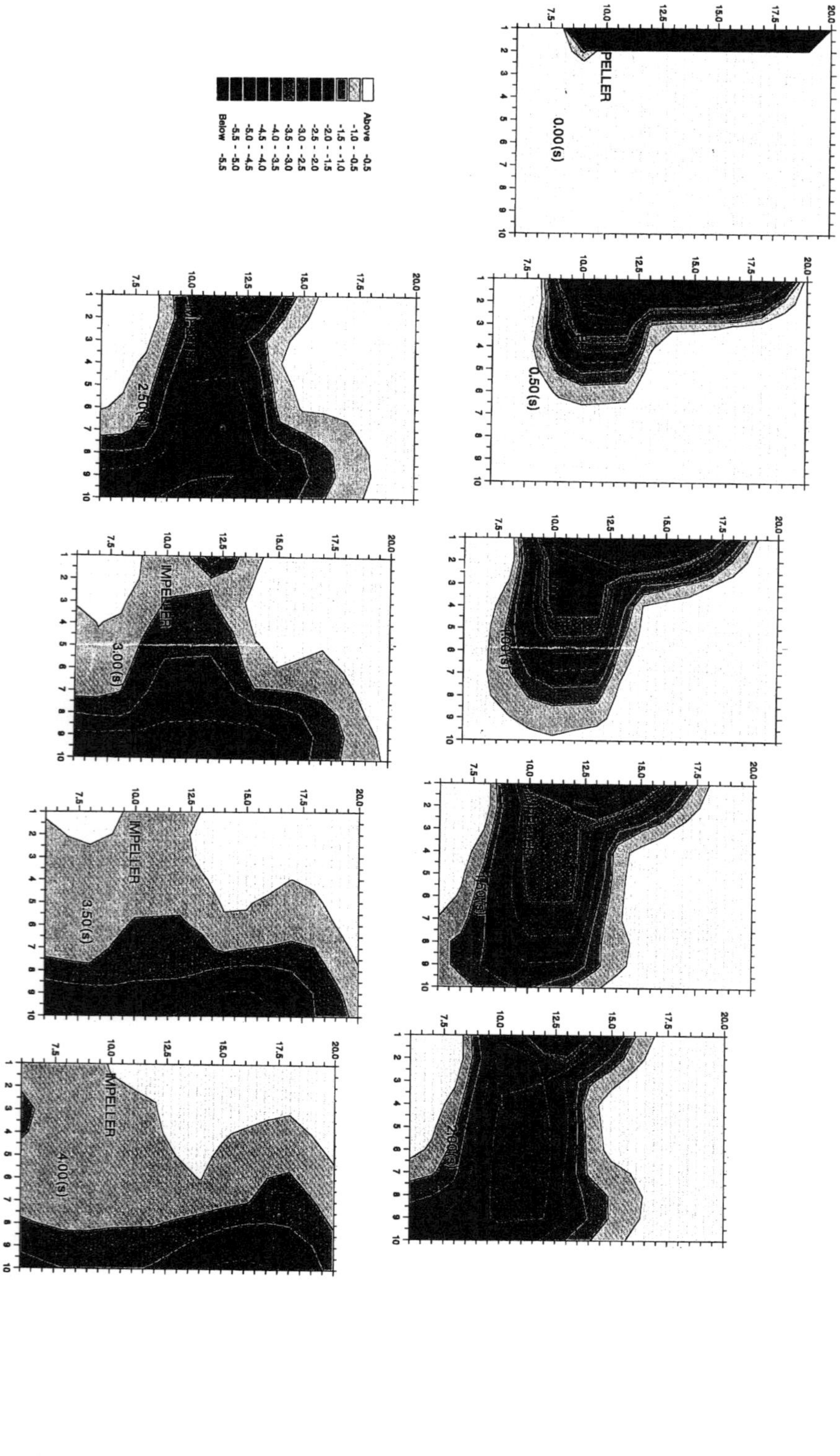

FIGURE 7 Simulation of Tracer Mixing by Networks-of-Zones

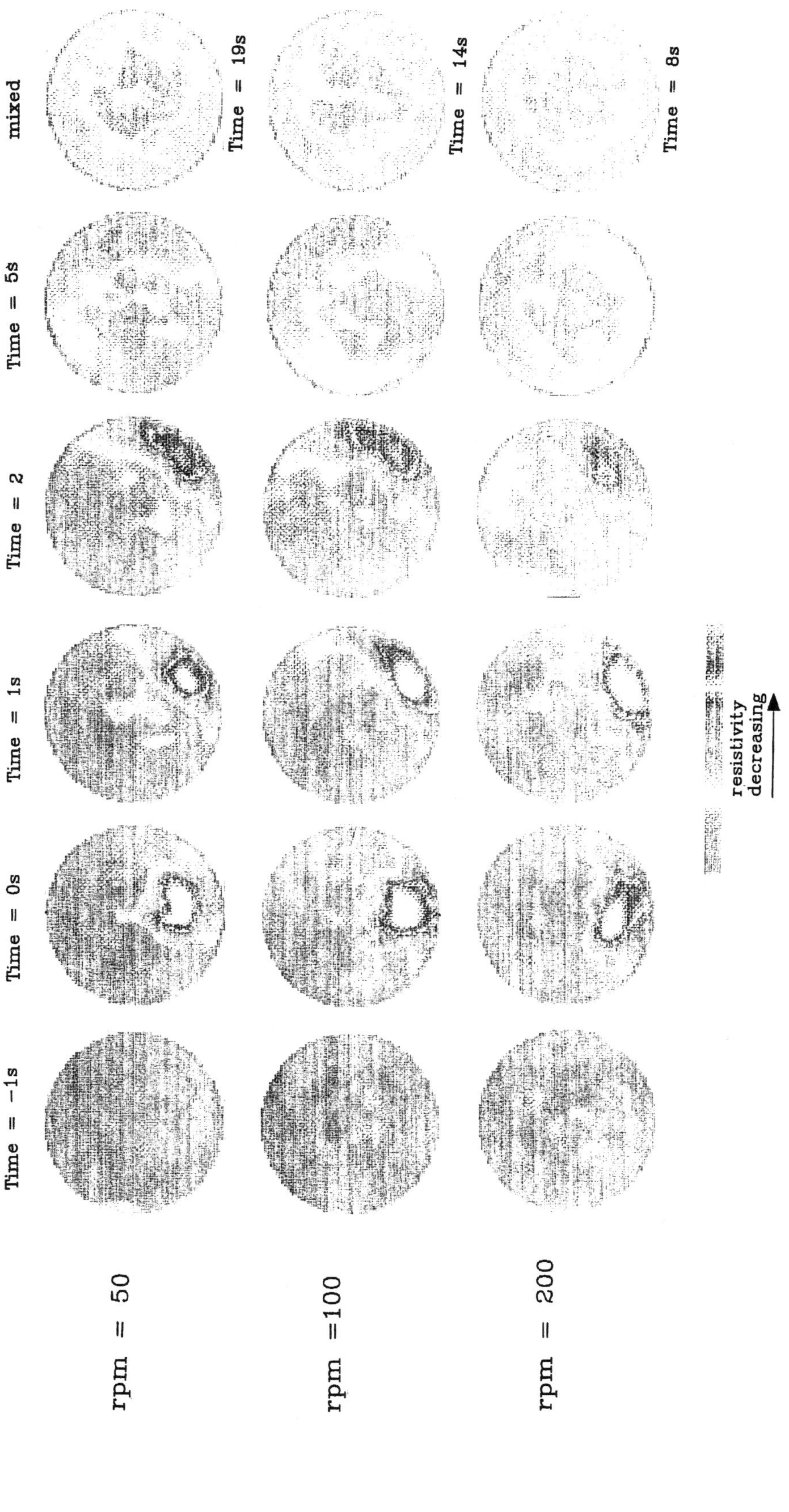

FIGURE 8 EIT Images for Asymmetric Single Point Tracer Injection

Mixing of Inert and Reactive Tracers in a Twin-Impeller Stirred Vessel

R. Mann and P. Ying

Chemical Engineering Department, UMIST, Manchester, U.K.

K. Baker

Crosfield Chemicals, Warrington, U.K.

R.B. Edwards

Unilever Research, Port Sunlight Laboratories, Merseyside, UK

Tracer mixing experiments in 'closed' batch stirred vessels are more difficult to interpret than the classical stimulus-response technique applied to continuous flow reactors, because concentrations vary in both space and time as mixing proceeds to completion. An extension of the use of a 2-D networks-of-zones approach to a twin-impeller stirred vessel is described. The use of thousands of zones gives a good spatial discrimination of internal mixing behavior. The mixing of a visible pulse of an inert tracer and the mixing and indicator visualized reaction of an alkali slug injected into a batch of acid have been captured on video. The visualization experiments are simulated and closely predicted by theoretical sectional image reconstructions of the spatial concentration fields evolving in time.

Introduction

Our understanding of mixing inside a batch stirred chemical reactor continues to be seriously underdeveloped. The mixing literature for chemical reactors has concentrated on the continuous flow case, using stimulus response methods. This technique relies on the injection of a tracer into an entering flow stream. The response to this stimulus is observed in the outlet flow. Batch and semi–batch reactors, which predominate in chemical manufacturing, cannot be evaluated in this way since there is no through flow. In these cases, tracer can be injected into a batch of stirred fluid, but the mixing process is then unsteady in space and time. If the response is measured at a single point (usually the impeller), a single timewise response is obtained, which is usually referred to as a mixing curve. The time taken to achieve some fractional approach to the final fully mixed concentration is the mixing time. Measurements at a single point need to be repeated at very many internal positions to build–up a reasonable picture of the spatial behaviour evolving in time.

Visualisation of tracer mixing inside a batch stirred vessel provides a simple and convenient method of realising space–time information on the character of the mixing process. Mixing of an inert visible dye tracer and acid/alkali tracer pairs made visible by suitable indicators can be captured on video. Selected frames can then be chosen to map out the mixing produced by convection and dispersion induced by stirrer rotation.

The network–of–zones model [1,2] presents a computationally convenient simplification of the full complexities of the circulatory turbulent mechanics. Moreover, networks–of–zones can very conveniently handle multiple reactions [3,4,5], which otherwise provoke difficulties of incorporation into the computational fluid dynamics (CFD) of turbulent flow [6]. The concept of the sub–division of a stirred vessel into networks–of–zones has also been shown to be applicable to the mixing of gas bubbles in a stirred gas–liquid vessel [7] as well as to the suspension of solids [8].

By using networks–of–zones, the assumption of perfect backmixing in a batch or semi–batch reactor can be dispensed with and the degree of partial segregation of reagents and intermediates directly predicted. Theoretical visualisations, for comparison with experiments, can be conveniently achieved by appropriate shading on sectional image reconstructions. These approaches have been extended to the case of a vessel fitted with a pair of Rushton turbine impellers in a semi–standard equal spacing configuration.

Network–of–Zones for a Pair of Impellers

The axi–symmetric arrangement of two 'square' networks–of–zones for a pair of impellers is shown schematically in Fig.1. Each impeller has two N^1 x N^1 networks [1,2] of backmixed zones. The total convection associated with an impeller, Q, is uniformly allocated to each of the N concentrically nested flow loops, so that each flow loop carries a flow $q = Q/2N^1$. The flow around each nested loop gives rise to backmixing in an axial circulating sense. The extent of this backmixing in a loop decreases as N^1 becomes larger. For very large N^1 the circulatory mixing approaches plug flow. The full set of nested flow loops accommodate the total convective flow produced by an impeller. Each backmixed zone undergoes an equal and opposite flow exchange with its lateral neighbours. This lateral mixing represents a turbulent mixing mechanism between parallel flowing streams. It is characterised by β which is the ratio of the exchange flow to the main loop flow. This turbulent mixing factor is taken to be constant, except within the impeller zones and those zones that connect the upper and lower circulations between the impellers, where an order of magnitude increase is applied [1]. The unsteady material balance for a component I for a single backmixed zone of volume $V_{i,j}$ is then simply given by the conservation ordinary differential equation of the form:

$$V_{i,j} \frac{dC_{I,i,j}}{dt} = q \left[C_{I\,i-1,j} - (1+2\beta)\, C_{I\,i,j} + \beta\, (C_{I\,i,j-1} + C_{I\,i,j+1}) \right] \pm r_{I_{i,j}} V_{i,j}$$

For an inert tracer $r_I=0$. For an acid (A) and alkali (B) reaction, the local rate of reaction is given by $kC_{A\,i,j}\, C_{B_{i,j}}$. Equation (1) then forms a set of $4N^{1^2}$ ordinary differential equations for each component I which can be conveniently integrated by NAG library subroutines. Examples will be predicted for $N^1=20$. A pair of impellers is then simulated by 1600 eqns for inert tracer mixing and 3200 for acid/alkali reaction. The sub–division of the fluid into 1600 equal volume zones is visualised in Fig.2, which achieves a reasonable level of spatial sub–division.

Inert Tracer Mixing

Fig.3 then shows a sequence of experimental video stills of the axisymmetric mixing of a pulse of inert nigrosine blue dye injected axisymmetrically from four syringes around the impeller shaft close to the upper fluid (water) surface. The stirrer speed was 100 rpm for a 22 litre charge. At t=0 the tips of the syringes are just visible, protruding below the surface. The experimental views are from a forward mid–point between the impellers. After 0.5 s the blue dye is being dispersed through the upper impeller. At 1.0 s, the tracer is splitting between the upper and lower circulation loops and has partially penetrated along the vessel walls forming a distinctive hammer–head outline. At 1.5 s the upper half vessel appears filled with dye and at 2.0 s it is beginning to be swept downwards towards the lower impeller. This progress towards the lower impeller becomes more advanced at 3, 4 and 5 s. Fig.2 also shows the networks–of–zones predictions as sectional image reconstructions using a line shaded zones technique. A single horizontal line per zone corresponds to $0.5 \leq C_I/C_{I\infty} \leq 1.5$, so that the final appearance at $C_I/C_{I\infty}$ would correspond to single line shading throughout the vessel. The maximum shading in Fig.2 is five lines per zone corresponding to $4.5 \leq C_I/C_{I\infty} \leq 5.5$. The predicted inert tracer dispersion closely follows the experimental visualisation. The parameters used were $Q=2.75\ ND^3$ for each impeller, with equal flow splits above and below each impeller and a $\beta=0.4$, which is somewhat higher than the previously reported value for a single impeller case [1,2].

Acid–Alkali ReactiveTracer Mixing

The experimental visualisation of an injected pulse of sodium hydroxide solution (10.0 M) into a batch of 3.7 x 10^{-3} M hydrochloric acid is again shown as a set of video stills in Fig.4. The alkali pulse injected is slightly less than that required to neutralise the batch of acid, so that the intense blue of the thymolpthalein blue indicator in alkali solution (pH $\geq$ 9.3) will eventually be reacted to a final clear solution. This indicator visualised case shows different behaviour from the inert tracer after the first second. There is a clear segregation of the fluid into an upper alkali (dark) region and a lower acid (clear) region, with some evidence of a partial, occasionally disordered, penetration towards the lower impeller. This behaviour is paralleled in the networks–of–zones simulation. There is an initial blue alkali penetration towards the lower impeller at 2.0 s. This is predicted to be followed by a temporary alkali 'retreat' into the upper half vessel at 3 s. This axisymmetric oscillation between the upper and lower circulation zones appears somewhat more ragged in the experimental case. This behaviour is thought to be associated with randomly meandering but essentially precessing eddies,which straddle the upper and lower radial flows and move with a tangential circulation. This effect cannot be achieved by a 2–D network–of–zones which uses only axisymmetric radial circulation. This third dimensional medium scale tangential flow is probably responsible for the somewhat faster reaction to decolouration in the experimental case. The network–of–zones predicts complete reaction of the deficit of acid after 18 s. However, it is emphasised that the 2–D theoretical predictions in Fig.4 provide a very close reproduction of the experimental visualisation, without resort to any parameter adjustments or further assumptions, except that the micro–mixing is assumed locally perfect for the reaction case.

Conclusion

It is therefore concluded that the network–of–zones, as extended for the twin–impeller case, provides a close simulation of the mixing of both inert and acid/alkali tracers, except that some visual evidence for asignificant unstable third dimensional tangential mixing may need incorporating into the model. Evidence for such an unstable mixing in the twin impeller case has just recently been reported [9].

Nomenclature

C_I	concentration of component I
$C_{I\infty}$	ultimate fully mixed concentration of component I
D	impeller diameter (Rushton turbine)
i	radial zone position
j	axial zone position
N	stirrer speed
N^1	network dimension
Q	total flow generated by impeller
q	flow in a circulation loop ($q=Q/N^1$)
r_I	rate of reaction for component I
t	time
$V_{i,j}$	volume of i, j zone
β	turbulent exchange parameter

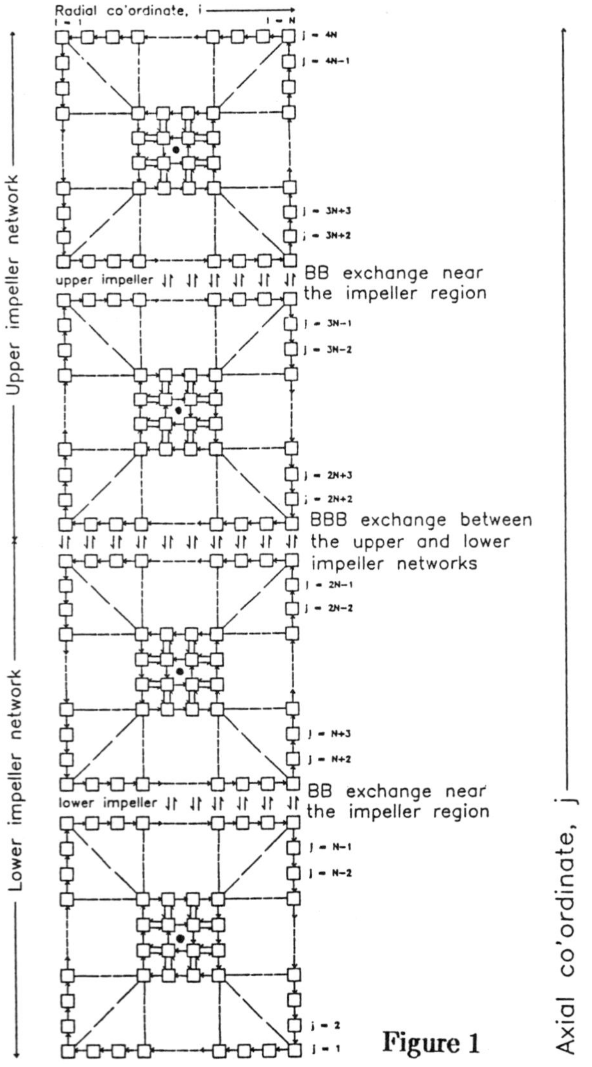

Figure 1

The 2–D Networks–of–Zones Configuration for a Twin–Impeller Stirred Vessel.

References

[1] Mann, R. and Knysh, P., I Chem E Symp. Series, 89 , 127, 1984.

[2] Mann, R., et al, I Chem E Symp. Series, 108, 55, 1987.

[3] Mann, R. and Wang, Y.D., I Chem E Symp. Series, 121, 241, 1990.

[4] Mann, R. and El–Hamouz, A.M., Proc. 7th Europ. Conf. Mixing, 1, 1991.

[5] Wang, Y.D and Mann, R., Trans. I Chem E, 70 (part A), 282, 1992.

[6] Pericleous, K.A. and Patel, A., Int. Jnl. P.C.H., 2, 105, 1986.

[7] Mann, R. and Hackett, L.A., Proc. 6th Europ. Conf. on Mixing, Pavia, 321, 1988.

[8] Brucato, A., Magelli, F., Nocentini, M. and Rizzuti, L., Trans I Chem E, 69 (part A), 43, 1991

[9] Mahmoudi, S.M. and Yianneskis, M., Proc. 7th Europ. Conf. Mixing, 17, 1991.

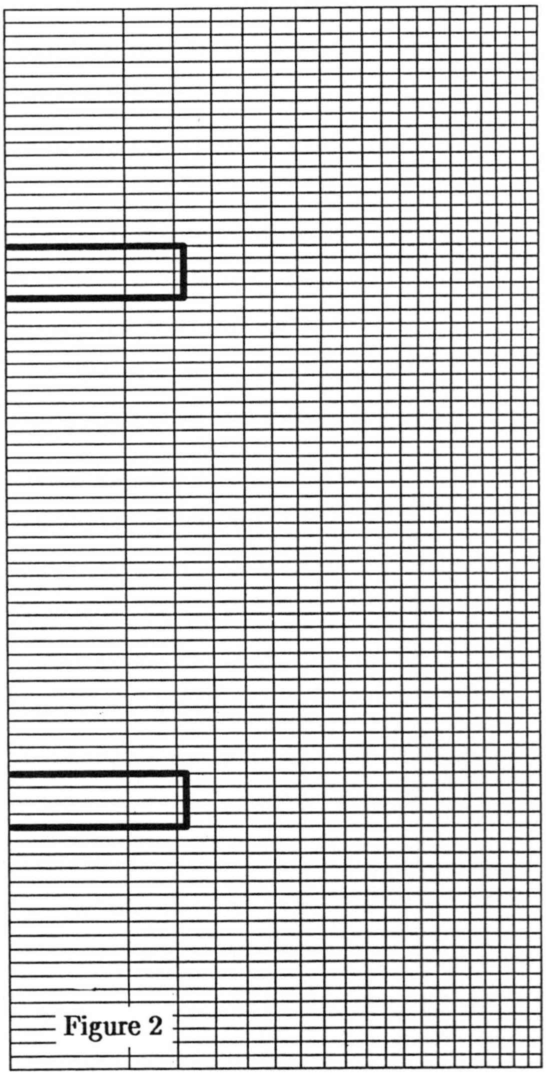

Figure 2

Disposition of 1600 Equal Volume Zones for 2 x (20 x 40) Networks

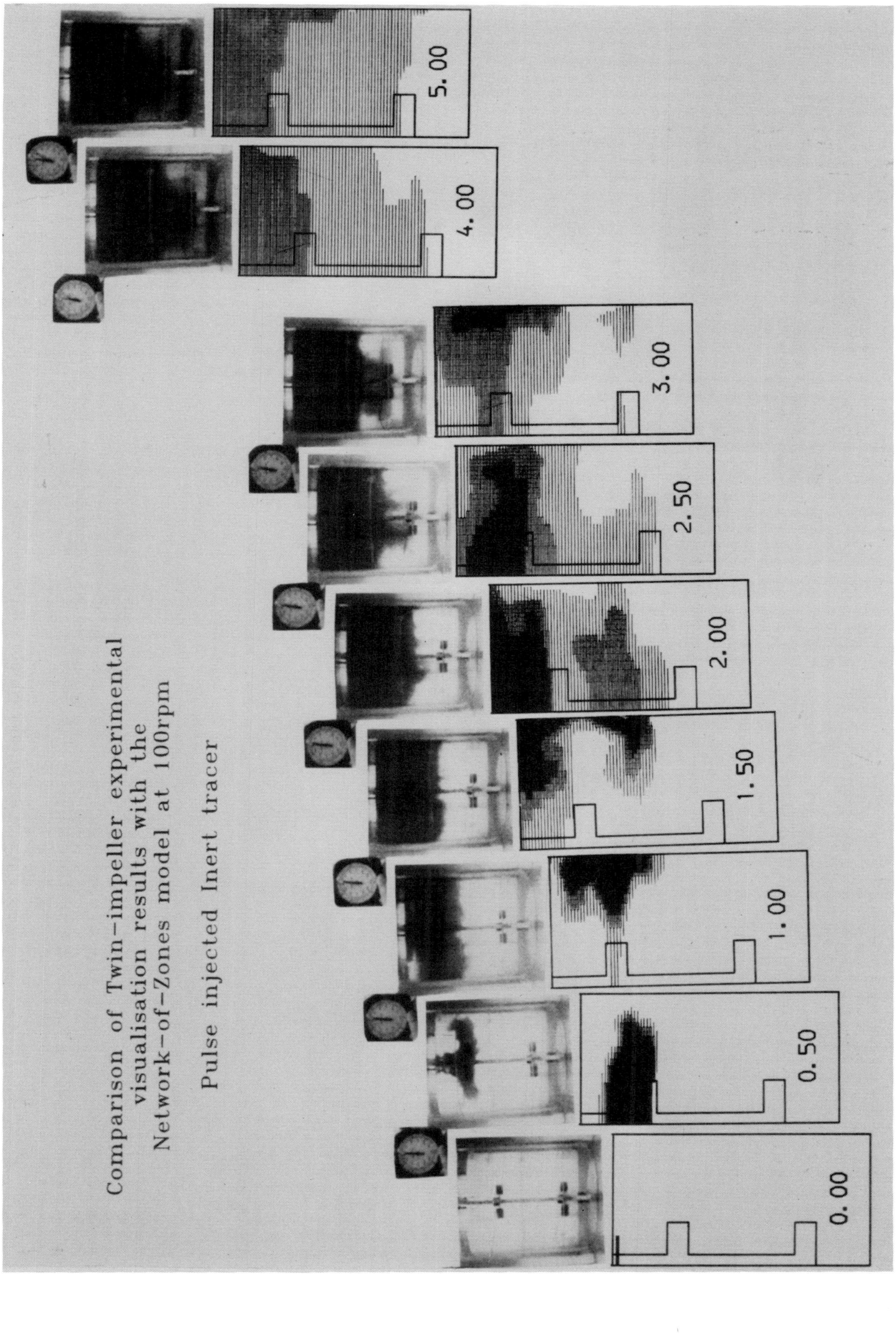

Figure 3 Inert Tracer Mixing for a Pulse Injection

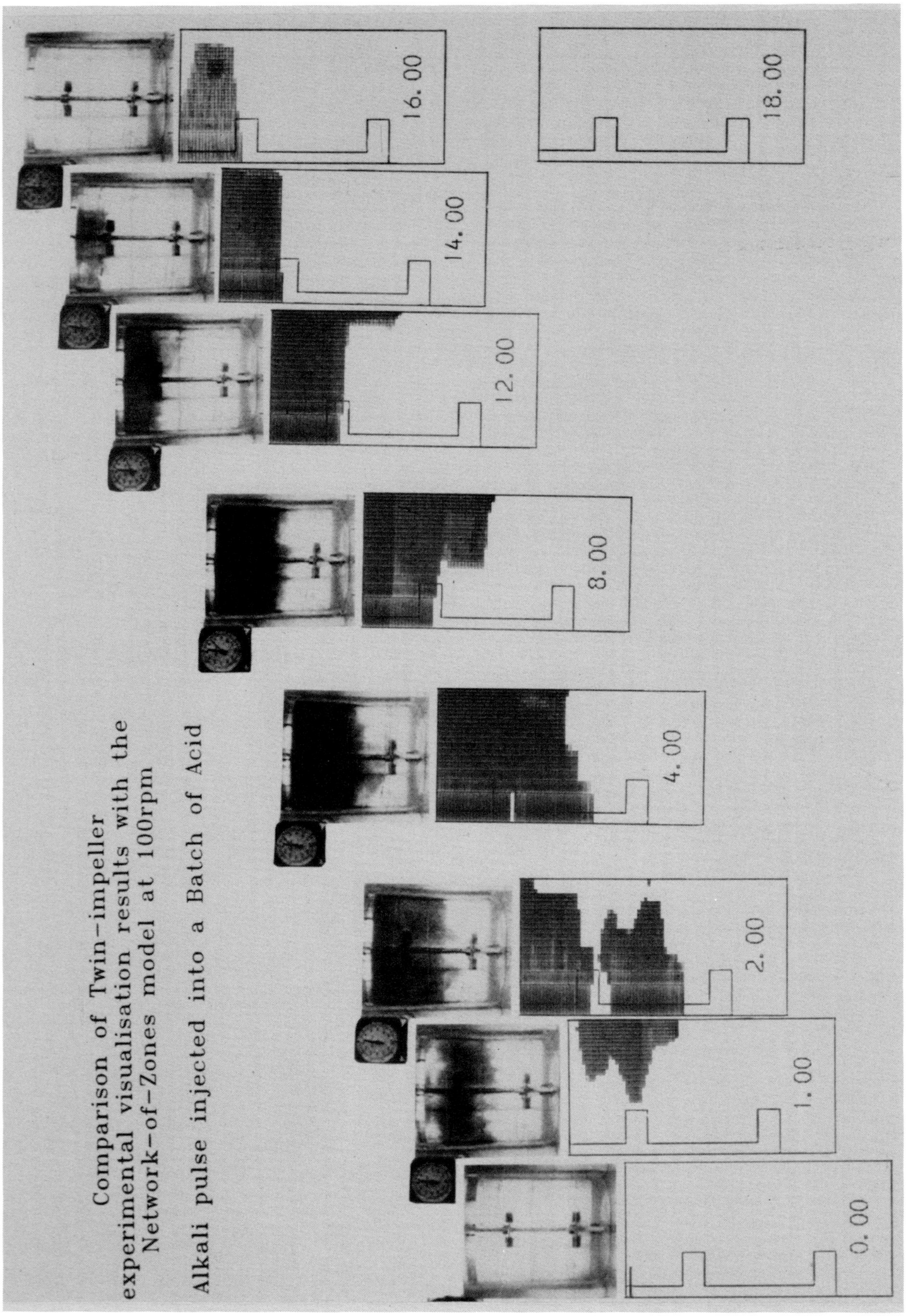

Figure 4 Reactive Tracers: Pulse of Alkali Injected into a Batch of Acid

An Interpretation of Experimental Micromixing Data for Tubular Reactors with a Three-Stage Micromixing Model

Massimo Pipino, Antonello A. Barresi, and Giancarlo Baldi

Dipartimento di Scienza dei Materiali e Ingegneria Chimica, Politecnico di Torino, C. so Duca degli Abruzzi 24, 10129 Torino, Italy

In order to take into account the turbulence effects on micromixing, a new approach is presented and applied to the derivation of a new correlation for the deformation of the fluid elements. This result is compared with existing analogous correlations [6, 8] and, used as a part of a three-stage micromixing model, is validated by experimental data from literature.

Mixing at molecular level (micromixing) can be an important step of the process of mixing and reaction of unpremixed reactant fluids. If very fast reactions occur, the characteristic time for micromixing can be larger than the characteristic reaction time and in this case mixing at molecular level becomes the controlling step of the process; moreover, yield and selectivity of the reactions are influenced by micromixing. Thus, for a wide range of applications, a micromixing model is necessary in order to predict the characteristic times (very fast reactions), the selectivity of the process (complex reactions), the product properties (precipitation and polymerization reactions) or to evaluate the stability of the reactor if runaway reactions may occur.

With the exception of very viscous fluids, the reaction media is usually turbulent. The mixing of the reactants is strongly dependent on the velocity field in the fluid: thus, a micromixing model must take into account the turbulence effects. If we refer to the spectral analysis of the concentration fluctuations, we can distinguish three main regions of the spectrum: the *inertial-convective range*, characterized by the convective motion of the lumps of matter, the *viscous-convective range*, in which the scale of segregation is reduced by the effect of viscosity and the *viscous-diffusive range*, where diffusion is the main mixing mechanism.

In previous works [1, 2] we proposed to model the effects of the different parts of the concentration spectrum, and the three main identified mixing mechanisms, in a three-stage micromixing model describing respectively (i) the *turbulent mixing*, (ii) the *deformation of fluid elements* and (iii) the *diffusion and chemical reaction*.

All the steps are influenced by turbulence effects, but the first one (turbulent mixing) is also dependent on the geometry of the apparatus in which the process takes place: a new approach [1] has been previously proposed to the description of this stage, in which macromixing and micromixing phenomena, like turbulent dispersion by large eddies and reduction of the segregation scale, are coupled. For the diffusion and chemical reaction stage, a lamellar model [3] has been adopted, but a new dimensionless form is suggested [2].

In this work, a new model for the deformation stage is proposed on the base of the spectral analysis of turbulence: to this aim, a new method for taking into account the turbulence effects is presented and applied to the calculation of the viscous deformations of a fluid element. The three-stage micromixing model has been modified accordingly and its predictions have been compared with experimental data from literature [4].

SPECTRAL ANALYSIS OF TURBULENCE

Only homogeneous isotropic turbulence in incompressible fluids will be considered in the following. It is always possible to express the three-dimensional energy spectrum $E(k)$ as the product of the Kolmogoroff spectrum law

$$E_{in}(k) = A\, \varepsilon^{2/3}\, k^{-5/3} \qquad (1)$$

(in which A=1.7) valid in the inertial subrange where the viscosity effects are not important, and a weight function $M(k)$ taking into account the dissipation terms:

$$E(k) = E_{in}(k)\; M(k) \qquad (2)$$

Referring to the energy spectrum proposed by Pao [5], it is possible to express the $M(k)$ function as:

$$M(k) = \exp\left(-\frac{3}{2}\, A\, (k\, \lambda_K)^{4/3}\right) \qquad (3)$$

in which λ_K is the Kolmogoroff microscale: $\lambda_K = (\nu^3/\varepsilon)^{1/4}$.

In the light of Pao's work, it is possible to show [6] that $M(k)$ represents the fraction of the total dissipation rate ε due to the eddies with wavenumber greater than k; Figure 1 shows the behaviour of the $M(k)$ function.

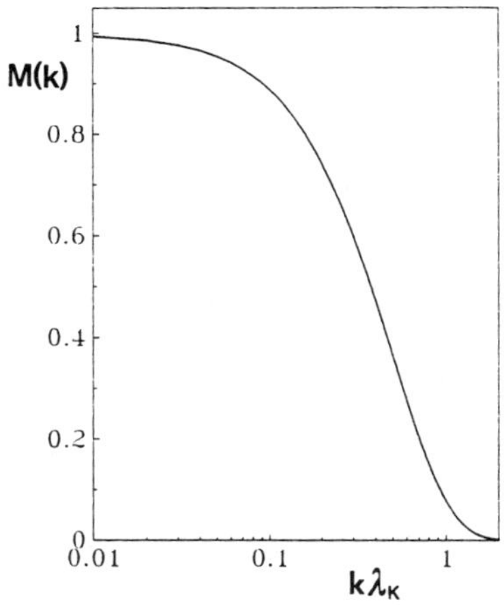

Figure 1. The $M(k)$ function.

But if we model the turbulent field as composed by eddies having a pure inertial behaviour and contributing to the spectrum according to a distribution function $M(k)$, then $M(k)$ in Equation (2) can be interpreted as the cumulative distribution function of wavenumbers greater than k in the total spectrum.

This interpretation may be very helpful in turbulence calculations: for example it is known [6] that the root mean square of the velocity of the eddies with wavenumber equal to k, $u'(k)$, can be defined as:

$$u'^2(k) = k\; E(k) \qquad (4)$$

In the inertial subrange, from Equation (1) it follows:

$$u'^2_{in}(k) = 1.7\; \varepsilon^{2/3}\; k^{-2/3} \qquad (5)$$

Therefore, considering Equation (2), we can pose in the whole spectrum:

$$u'^2(k) = u'^2_{in}(k)\; M(k) \qquad (6)$$

In order to obtain the characteristic turbulent velocity u' as a function of ε and ν, we can write:

$$u'^2 = \int_0^\infty u'^2_{in}(k) \cdot \frac{dM(k)}{dk}\, dk \qquad (7)$$

Equation (7) allows the evaluation of the root mean square of the velocity fluctuations from their spectral expression derived in the inertial subrange.

We suggest to use the same method adopted in Equation (7) for the determination of turbulent energetic quantities from their spectral analysis in the simplified case of pure inertial behaviour [7].

DEFORMATION OF THE FLUID ELEMENTS

A fluid element in a turbulent field is deformed by the velocity fluctuations of the fluid. Different correlations that describe this deformation have been presented in literature [6, 8]; in the present work the lamina-deformation hypothesis proposed by Baldyga and Bourne [8] is adopted. They considered the deformation of a lamella of initial thickness d_0 which, under the effect of the turbulent motion, shrinks in one direction and grows up in another one, keeping constant its third dimension. Baldyga and Bourne took into account the turbulent effects by adopting the velocity-difference correlation (see [6], pg. 406):

$$\mathbf{W} = \overline{w_i(d_0)\, w_i(d_0)} \qquad (8)$$

where $w_i(d_0)$ are the components of the relative velocities of pairs of fluid particles initially distant d_0 from each other.

The Baldyga and Bourne approach leads to Equation (9), that describes the behaviour in time of the shrinking lamellae:

$$\left(\frac{d}{d_0}\right)^2 = 1 + \frac{3\mathbf{W}}{2d_0^2}t^2 - \left(\left(1 + \frac{3\mathbf{W}}{2d_0^2}t^2\right)^2 - 1\right)^{1/2} \qquad (9)$$

Baldyga and Bourne posed $d_0 = \lambda_K$ and used for $\mathbf{W}$ the correlation obtained by Hinze [6] for viscous dissipation motion, deriving for $\mathbf{W}/d_0^2$ the expression:

$$\frac{\mathbf{W}}{d_0^2} = \frac{1}{3}\left(\frac{\varepsilon}{\nu}\right) \qquad (10)$$

If we note that the Kolmogoroff microscale λ_K is defined as the scale at which the convective transport balances the viscous transport, it follows that Equation (10) should be valid only if d_0 is negligible in respect of the Kolmogoroff microscale.

We propose to introduce a spectral analysis of the correlation $\mathbf{W}$: we know [6] that at very high k-values $\mathbf{W} = (\varepsilon/3\nu)\, d_0^2$ and that at low wavenumber values

$$\mathbf{W} = 8.25\ \varepsilon^{2/3}\ d_0^{2/3} \qquad (11)$$

An eddy associated with the wavenumber k is characterized by the contemporaneous presence of convective and viscous behaviours; their relative importance might be estimated referring to the eddy Reynolds number Re_k that we define as:

$$Re_k = \frac{\lambda\ u^*}{\nu} \qquad (12)$$

in which $\lambda = 1/k$ is the length scale of the eddy and

$$u^* = \varepsilon^{1/3}\ k^{-1/3} \qquad (13)$$

is the velocity scale of the eddy derived by dimensional analysis. This choice for the expressions of λ and u^* makes possible to write, from Equations (12) and (13):

$$Re_k = (k\ \lambda_K)^{-4/3} \qquad (14)$$

In this way, the eddy Reynolds number in correspondence of the Kolmogoroff microscale is equal to unity.

As Re_k is the ratio between inertial and viscous transport, it can be used as a weight function in order to interpolate the limiting Equations (10) and (11):

$$\mathbf{W}(k) = \left(\frac{\varepsilon}{3\nu}\ d_0^2\right)\frac{1}{1+Re_k} + (8.25\ \varepsilon^{2/3}\ d_0^{2/3})\ \frac{Re_k}{1+Re_k} \qquad (15)$$

Equation (15) evidently follows Equation (10) for very high k-values (Re_k goes to zero) and equation (11) for very low k-values (Re_k goes to infinity).

If we consider that an eddy associated to the wavenumber k has a characteristic dimension s_0 [9]:

$$s_0 = 2\pi/k \qquad (16)$$

and that the deformation of a lump of matter follows the deformation of the eddy in which it is contained, we can substitute in Equation (15) the dimension d_0 with the expression for s_0. From Equations (15) and (16), we obtain:

$$\boldsymbol{W}(k) = 41.25 \left(\frac{\varepsilon}{k}\right)^{2/3} \frac{Re_k}{1+Re_k} \tag{17}$$

and

$$\frac{\boldsymbol{W}}{d_0^2}(k) = 1.05\ \varepsilon^{2/3}\ k^{4/3}\ \frac{Re_k}{1+Re_k} \tag{18}$$

An averaged value of $\boldsymbol{W}/d_0^2$ to be inserted in Equation (9) may be obtained by the method previously discussed, similarly to what shown in Equation (7):

$$\frac{\boldsymbol{W}}{d_0^2} = 0.19 \left(\frac{\varepsilon}{\nu}\right) \tag{19}$$

RESULTS AND DISCUSSION

The value of $\boldsymbol{W}/d_0^2$ obtained above is lower than the one proposed by Baldyga and Bourne [8] and given by Equation (10): this fact was expectable because we have considered the influence of all the eddies on all the scales while the correlation proposed in literature was valid only for d_0 negligible in respect of the Kolmogoroff microscale and was extended to a very wider range. In Figure 2 the thinning of a lamella calculated according to Equations (10) and (19) is compared.

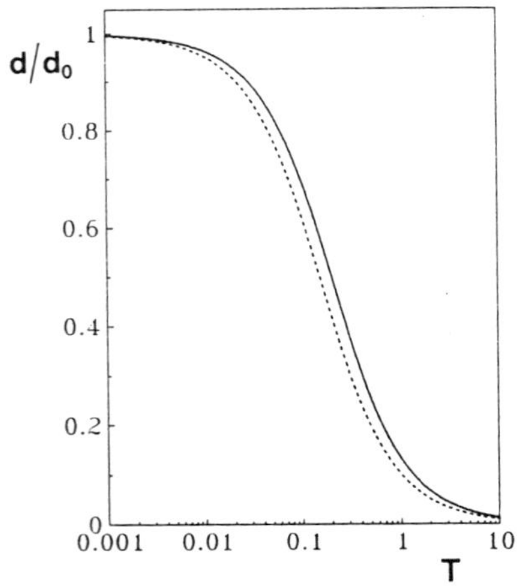

Figure 2. Comparison between the correlations for the deformation of the lamellae: (-----) this work; (- - -) Baldyga and Bourne [8].

The micromixing model previously proposed [1, 2] has been modified taking into account Equations (9) and (19), and the three mixing stages in a tubular reactor with isokinetic coaxial feeds are now described by the equations:

a) turbulent mixing time:

$$t_{TM} = 0.301\ d_i\ (\nu\ \varepsilon)^{-1/4} \tag{20}$$

b) deformation of fluid elements:

$$\left(\frac{d}{d_0}\right)^2 = 1 + 0.285\ \pi^4 T^2 - \left((1 + 0.285\ \pi^4 T^2)^2 - 1\right)^{1/2} \tag{21}$$

which has been derived by substituting Equation (19) in Equation (9). The dimensionless time:

$$T = (\varepsilon/\nu)^{1/2}\ t/\pi^2 \tag{22}$$

has been introduced.

c) diffusion and chemical reaction:

$$\frac{\partial C_i}{\partial T} = \frac{1}{Sc}\left(\frac{d_0}{d}\right)^2 \frac{\partial^2 C_i}{\partial X^2} - M\left(\frac{c_{A0}}{c_{i0}}\right) C_A C_B \tag{23}$$

in which we introduced the non-dimensional terms:

$$C_i = c_i/c_{i0} \tag{24a}$$

$$X = 2x/d_0 \tag{24b}$$

$$M = (\nu/\varepsilon)^{1/2}\ \pi^2\ c_{B0}\ k_r \tag{24c}$$

and where A and B are the species involved in the chemical reaction, Sc is the Schmidt number, x is the space coordinate along the width of the lamellae and k_r is the kinetic constant of the chemical reaction between A and B; further details can be found in [1, 2].

A tubular reactor with coaxial isokinetic feeds (see Figure 3) in

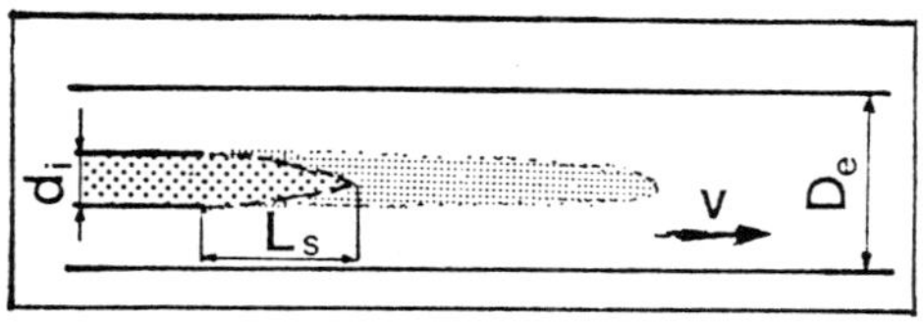

Figure 3. The scheme of the tubular reactor.

which an acid-base chemical reaction takes place has been considered; a detailed description of the experimental apparatus can be found in [2] and [4].

At the exit of the inner tube, it is possible to observe a colored plume, structured in two main parts: a highly segregated colored cone (of length L_s), which corresponds to the turbulent mixing stage, and a zone of vanishing color, in which deformation, diffusion and chemical reaction take place.

Some experimental data can be found in literature regarding this kind of apparatus: the length of the segregated zone (L_s) and of the whole plume has been optically measured; the predictions of the proposed model have been compared with the experimental data of Pohorecki and Baldyga [4], and a good agreement has been obtained (see Figure 4).

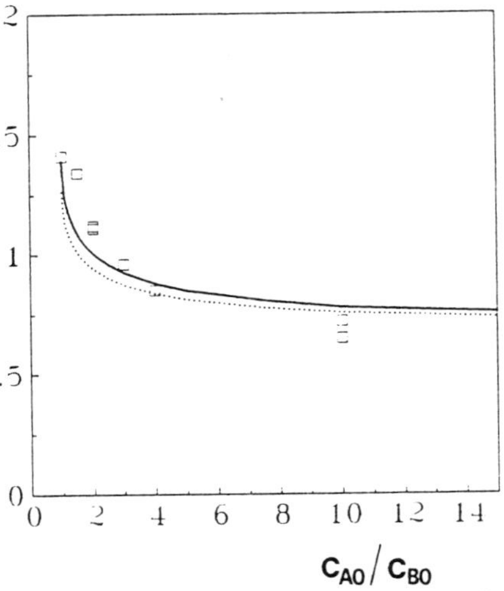

Figure 4. Comparison between (-----) the proposed model, (.....) the previous version (see [1] and [2]) and the experimental data from [4]. Re=13,000; d_i=5.2·10^{-3} m; D_e=0.04 m.

CONCLUSIONS

A new method for taking into account the turbulent effects has been proposed and applied to the calculation of the turbulent velocity-difference correlation in order to obtain a new formulation for the kinematics of a shrinking lamella. In this way both inertial and viscous effects of the eddies are accounted for. A refined version of the three-stage micromixing model previously proposed has been obtained, whose predictions are in good agreement with experimental data from literature.

ACKNOWLEDGMENTS

This work was financially supported by "Ministero dell'Università e della Ricerca Scientifica e Tecnologica" (40% MURST - Fluidodinamica multifase).

NOTATION

A	Kolmogoroff spectrum constant
C	non-dimensional concentration (see eq. 24a), -
c	species concentration, kmol m^{-3}
d	lamina thickness, m
D_e	tubular reactor diameter, m
d_i	inner tube diameter, m
$E(k)$	three-dimensional turbulent energy spectrum, m^3 s^{-2}
k	wavenumber, m^{-1}
k_r	kinetic constant, m^3 s^{-1} kmol^{-1}
L_s	length of the segregated zone, m
M	mixing modulus (see eq. 24c), -
$M(k)$	cumulative eddy distribution function, -
Re_k	eddy Reynolds number, -
Sc	Schmidt number, -
s	distance between pairs of fluid particles, m
T	non-dimensional time (see eq. 22), -
t	time, s
t_{TM}	turbulent mixing time, s
u'	turbulent velocity, m s^{-1}
$u'(k)$	eddy velocity, m s^{-1}
u^*	eddy velocity scale, m s^{-1}

W turbulent velocity-difference correlation, $m^2\ s^{-2}$

W(k) turbulent velocity difference correlation of the eddy of wavenumber k, $m^2\ s^{-2}$

$w_i(s)$ relative velocity of two fluid particles, $m\ s^{-1}$

X non-dimensional coordinate (see eq. 24b), -

x lamella axial coordinate, m

Greek letters

ε dissipated power per unit mass, $m^2\ s^{-3}$

λ turbulent scale, m

ν kinematic viscosity, $m^2\ s^{-1}$

Subscripts

0 initial conditions

A,B,i chemical species

in inertial subrange

K Kolmogoroff microscale

LITERATURE CITED

1. Pipino, M., Barresi, A.A. and G. Baldi, "Micromixing in tubular reactors: a new approach to the description of the 'decay of segregation' stage", in *Recents progres en genie des procedes: Agitation industrielle*, Vol 6-92-Nr 19, III-55, Lavoisier, Paris (1992).

2. Barresi, A.A., Pipino, M. and G. Baldi, *Chem. Eng. Sci.*, **47**, 9-11, 2831 (1992).

3. Angst, W., Bourne, J.R. and R.N. Sharma, *Chem. Eng. Sci.* **37**, 4, 585 (1982).

4. Pohorecki, R. and J. Baldyga, *Ind. Eng. Chem. Fundam.* **22**, 392 (1983).

5. Pao, Y.H., *Phys. Fluids* **8**, 1063 (1965).

6. Hinze, J.O., *Turbulence*, 2nd ed., McGraw-Hill, New York (1975).

7. Pipino, M., "Turbulence effects on micromixing", presented at Chem. Eng. Forum for Young Scientists/ISCRE 12, Torino, Italy (1992).

8. Baldyga, J. and J.R. Bourne, *Chem. Eng. Sci.* **39**, 329 (1984).

9. Tennekes, J.H. and I.L. Lumley, *A first course in turbulence*, MIT Press, London (1972).

Mixing Performance of Lattice-Type Twisting Blade for New Batch-Type Polymerization Reactor for High-Viscosity Liquid

Hidekazu Nakamoto and Oda Chikao

Mech. Eng. Research Lab., Hitachi Ltd., 794 Higashi-toyoi Kudamatsu-shi Yamaguchi, 744, Japan

Kazuo Ihara

Kasado Works, Hitachi Ltd., 794 Higashi-toyoi Kudamatsu-shi Yamaguchi, 744, Japan

We have investigated mixing performace of high viscosity liquid as high as several kPa•s in our original agitator, named the lattice-type twisting blade. This paper compares the mixing performance of the new blade with the other impellers. To do so, we measure the concentration of tracer in high-viscosity liquids, then estimate mixing time from the change in the standard devation of the tracer concentration. We use this method to study the relationship between mixing time and liquid viscosity for each impeller. Finally, using a scale-up of the new blade, we investigate the effects of blade size and the relationship between mixing performance and the liquid volume.

INTRODUCTION

Plastics research has been very active recently, improving existing plastics and developing new ones. These plastics are for many applications and are usually produced in small quantities by batch-type polymerization reactors handling treating liquid with viscosities as high as several thousands Pa·s. In the standard reactor, high-viscosity treating liquid adheres to and remains on the surface of the shaft or blade, decreasing mixing performance. This results in the deterioration of quality and productivity for long-term production. To solve these problems, we have developed a new type of blade which has no shaft and boasts a unique structure. Here, we report the results of investigation on its mixing performance in treating high-viscosity liquid. For low-viscosity liquid treatment, several methods [1,2,3] have been applied to estimate the mixing performance. But these are not applicable for highly viscous liquids because of the liquids' small fluidity and they disturb light transmittance to mix with the circumference gas. Therefore, we estimate the mixing performance using the standard deviation of the tracer concentration. We have also studied a scale-up of the new blade.

EXPERIMENTAL

NEW TYPE BLADE

Figure 1 shows the geometrical configuration and a photograph of the new blade. This agitator, named the lattice-type twisting blade, has no shaft which contacts the treating liquid. It is constructed of three frame pieces. Each frame piece comprises two sticks and two thick plates. The thick plate member curves along the surface of the container wall. When this agitator begins to rotate, the thick plates scrape the treatment for the highly viscous liquid upward. The entire surface of the container wall is scraped, minimizing dead space.

STANDARD IMPELLER

Figures 2 and 3 show the geometrical configurations, of the impellers used for the mixing

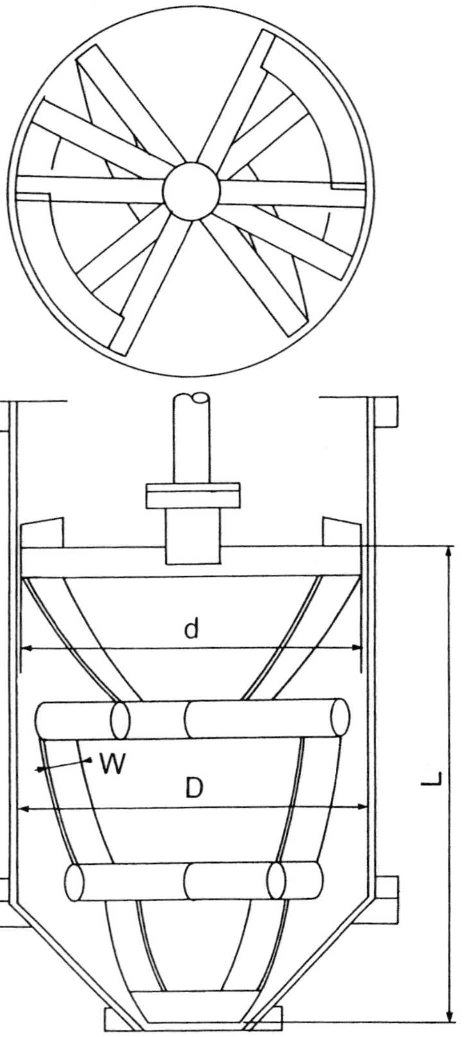

D=390[mm] L=530[mm] d=380[mm] W=40[mm]

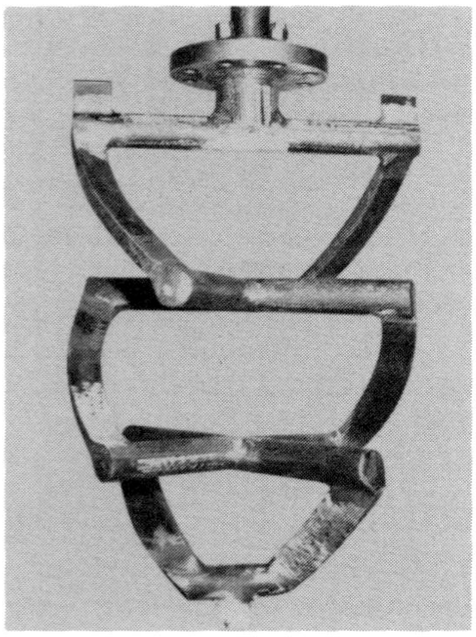

Figure 1. Geometrical configuration and photograph of new blade

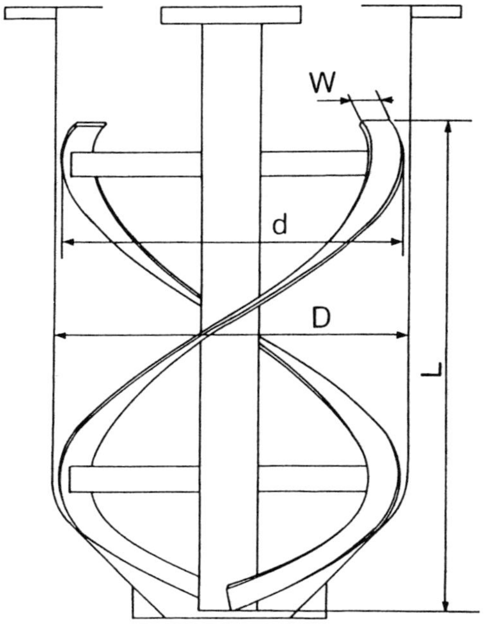

D=390[mm] L=530[mm] d=380[mm] W=40[mm]

Figure 2. Geometrical configuration of double-helical ribbon impeller

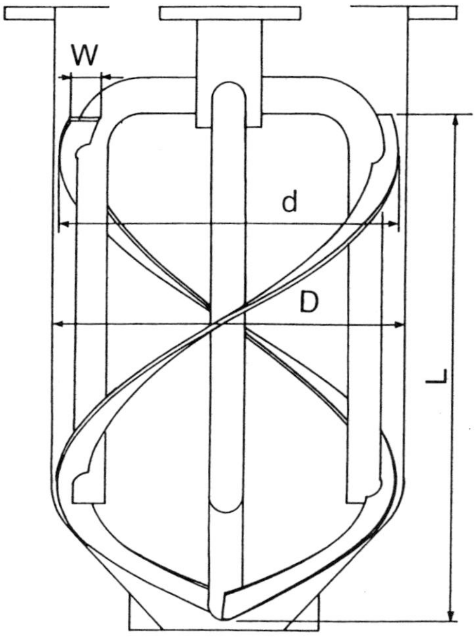

D=390[mm] L=530[mm] d=380[mm] W=40[mm]

Figure 3. Geometrical configuration of shaftless double-helical ribbon impeller

performance comparison, the double-helical ribbon impeller in

Figure 2 and the shaftless type double-helical ribbon impeller in Figure 3. These impellers' diameter, height and width are of the same scale as in the lattice-type twisting blade shown in Figure 1.

EXPERIMENTAL METHOD AND ESTIMATION METHOD

As the highly viscous liquid, corn syrup with various viscosities was used for tests. The viscosity regulates the hydrous rate of corn syrup. Mixing performance was judged from the tracer concentration change. First a small amount of tracer (3.0% conc. nitrate metal, 0.2 wt% of total treating liquid) with viscosity equal to that of the treating liquid was poured onto the surface of the liquid in the container. After each trial, of the required term and rotating speed, samples were collected from various points in the container and their tracer concentration were measured. From these concentration analyses, we calculated the standard deviation using the final average concentration of the mixing container and the final average concentration for all sampling times. We determined the ideal number of uniform mixing revolutions N·θ (N: rotation speed, θ: rotational time) from the pattern of convergence for these standard deviations. We investigated the effect on N·θ of changes in the treating liquid viscosity. In scale-up observations, we determined N·θ for two kinds of container and four kinds of blade. The specifications for the experimental impellers are shown here in Table 1.

TEST RESULTS AND DISCUSSION

NUMBER OF REVOLUTIONS FOR UNIFORM MIXING

Figure 4 shows the distribution concentration of tracer in the container. As the mixing revolutions increase, the bands for tracer concentration become narrower, becoming one constant value after a certain number of revolutions. It clearly appears two ranges where the standard deviation becomes suddenly small and does not change very much above a certain value. We have determined this point (an intersection of these ranges) for the number of uniform mixing revolutions,and plot it against viscosity in Figure 5. It is seen that the viscosity of the mixing liquid increases faster than N·θ with this new type of blade. But for the other impellers, N·θ strongly depends on the viscosity of the mixing liquid. When the

Table 1. Geometric variables of new blade

Geometry No.	D [mm]	d [mm]	L [mm]	W [mm]
1	710	700	1150	74
2	710	691	1150	74
3	390	380	530	40
4	390	370	530	40

D : Container diameter d : Blade diameter
L : Length of blade W : Width of blade

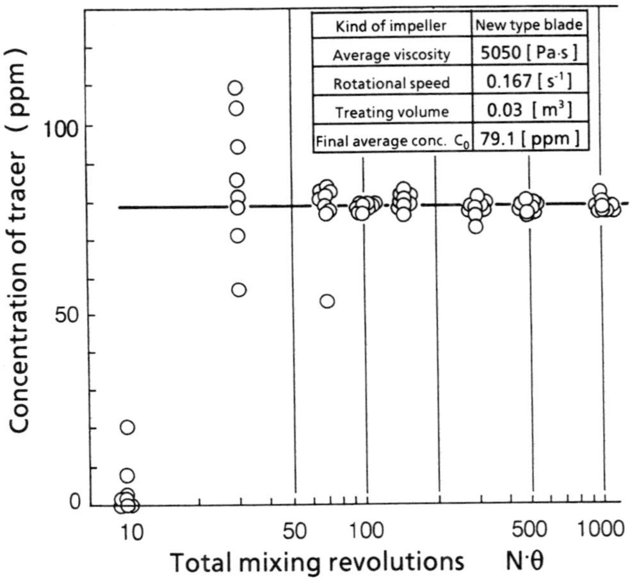

Figure 4. Tracer concentration vs. N·θ

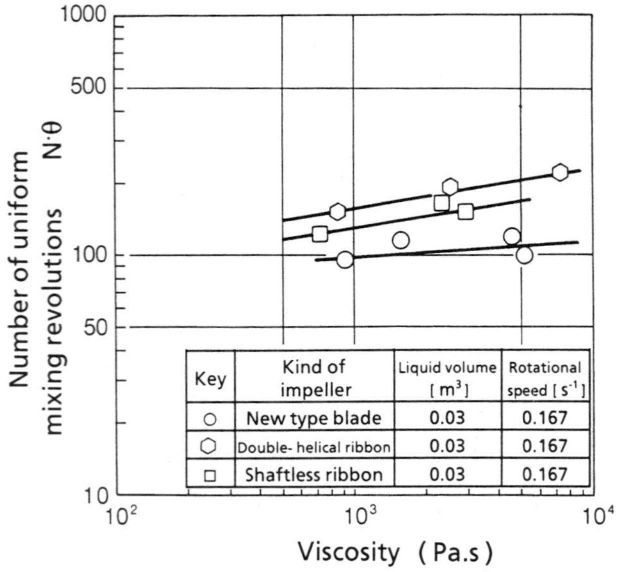

Figure 5. Mixing performance (1)

mixing liquid viscosity is 5 kPa.s, the lattice-type twisting blade N·θ is 110 while the other impellers' N·θs are from 160 to 200. From this, it is clear that the new blade has excellent mixing performance for highly viscous liquids.

SCALE-UP

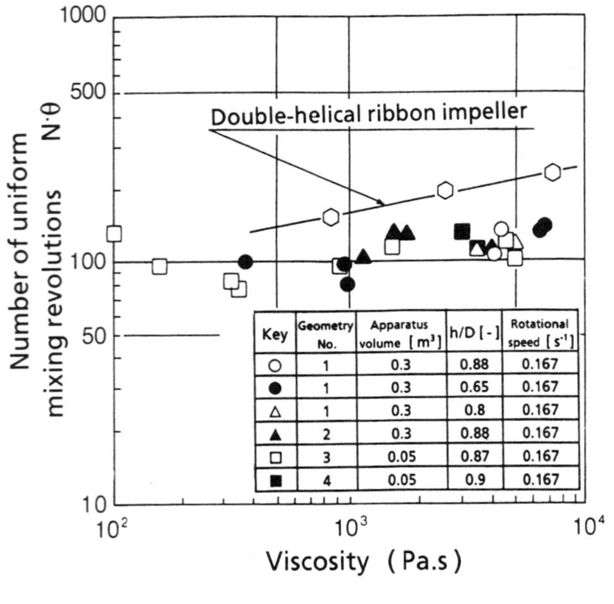

Figure 6. Mixing performance (2)

Figure 6 shows N.θ vs. viscosity as a function of blade size, treating volume and the blade gap using containers of 0.05 m^3 and 0.3 m^3. The figure show that in the region of mixing viscosity from 0.1 kPa·s to 7 kPa·s, the new blade has a tendency to slightly increase N.θ as the mixing liquid viscosity increases. The equipment size and volume of treating liquid are little affected in the region of volume h/D (h: height of liquid surface, D: diameter of container) 0.65 to 0.9 .

CONCLUSION

The new blade shows mixing performance for highly viscous liquids clearly superior to the standard impellers in the region of mixing viscosity 0.1 kPa·s to 7 kPa·s. Also to measure the concentration of tracer becomes a very useful estimation method to determine the mixing performance for highly viscous liquids.

LITERATURE CITED

1. J. B. Gray: Chem. Eng. Prog. , 59, 55 (1963).

2. M. Kamiwano and M. Hasegawa: Kagaku Kougaku Ronbunshu, 5, 65 (1979).

3. M. Kamiwano and K. Ohya: Kagaku Kougaku Ronbunshu, 5, 105 (1979).

Mixing Performance Analysis for an Intermeshing Counterrotating Twin Screw Extruder

Tao Li and Ica Manas-Zloczower

Dept. of Macromolecular Science, Case Western Reserve University, Cleveland, OH 44106

Three-dimensional flow patterns of a power-law model fluid in the regions of conveying elements and shearing discs of a Leistritz LSM30.34 intermeshing counterrotating twin screw extruder were calculated. The dispersive mixing efficiency was characterized in terms of shear stresses generated and a parameter λ quantifying the elongational flow components. The distributive mixing process in the C-chamber was studied by means of particle trajectories and the evolution of a pairwise correlation function. A parameter ε was used to quantify the goodness of distributive mixing.

The intermeshing counter-rotating twin screw extruder is widely used for dispersive mixing applications, devolatilization and reactive extrusion. The extruder contains various modular elements which provide different functions. In the conveying elements the material is transported to the die end of the extruder by positive displacement. The leakage flows occurring in the intermeshing regions, although reducing the degree of positive conveying, enhance the mixing features of the conveying elements. The shearing discs are non-forwarding elements which intensify mixing.

Various research groups have developed analytic models for the flow within the C-chamber and the leakage flows for Newtonian fluids(1-7). Numerical simulations for the 2-D flow in the calender gap for both Newtonian and power law fluids were also reported (8). Here we present a mixing performance analysis based on the 3-D flow patterns for a power law model fluid. A fluid dynamics analysis package-FIDAP(9), using the finite element method was employed to simulate the flow patterns in the regions of conveying elements and shearing discs. The dispersive mixing efficiency was characterized in terms of shear stresses generated and a parameter λ quantifying the elongational flow components. Distributive mixing in the C-chamber was studied by means of a pairwise correlation function.

CALCULATION OF THE FLOW FIELD

The simulations were carried out for a Leistritz LSM30.34 intermeshing counterrotating twin screw extruder. The equations of continuity and motion for the steady-state, isothermal flow of an incompressible fluid in the regions of conveying elements and shearing discs were solved:

$$\underline{\nabla} \cdot \underline{v} = 0 \qquad (1)$$

$$\underline{\nabla} \cdot \rho \underline{v}\,\underline{v} = -\underline{\nabla} P - \underline{\nabla}\,\underline{\underline{\tau}} \qquad (2)$$

The material used was HDPE, whose rheological behavior can be described by a power-law model with a Newtonian plateau:

$$\underline{\underline{\tau}} = \begin{cases} -\mu \underline{\underline{\dot{\gamma}}} & (|\dot{\gamma}| \le \dot{\gamma}_0) \\ -m|\dot{\gamma}|^{n-1} \underline{\underline{\dot{\gamma}}} & (|\dot{\gamma}| > \dot{\gamma}_0) \end{cases} \quad (3)$$

where the consistency index $m=4680\ N{\cdot}s^{0.59}/m^2$, the power law index $n=0.59$, and the Newtonian viscosity $\mu=1160.45\ N{\cdot}s/m^2$.

The mesh designs are shown in Figure 1. The elements used in the finite element method were brick elements with 8 nodal points in each element. In the conveying elements region the total number of nodal points is 27107 and the total number of elements is 22344. There are 14928 elements and 19110 nodal points in the shearing discs region.

The no-slip boundary conditions on the screw surfaces and barrel walls were used. For the conveying element region, this type of boundary condition covers most part of the boundary surfaces. Only the conditions at the outside cross-section of the clearance gaps for the first and third segments were not specified. Instead we used a nominal value for the normal stress difference in the axial direction and an iterative procedure based on symmetry analysis to calculate the tangential pressure distributions. In the region of the shearing discs, a nominal value for the normal stress difference in the axial direction was used as the boundary condition.

All the flow patterns were calculated by using FIDAP on a CRAY-YMP supercomputer. The cpu time required to perform the flow simulations varied between 6000 to 15000 sec depending on flow conditions.

DISPERSIVE MIXING EFFICIENCY

Experimental observations by Elmendorp (10) and the theoretical results by Manas-Zloczower and Feke(11,12) show that elongational flows are more effective than simple shear flows for dispersive mixing. The shear stresses generated in the flow field and a parameter λ quantifying the elongational flow components were used to characterize the flow field. The parameter λ is defined as:

$$\lambda = \frac{\dot{\gamma}}{\dot{\gamma}+\omega} \quad (4)$$

where $\dot{\gamma}$ and ω are the magnitude of the rate of deformation tensor and the vorticity tensor, respectively. The values of λ vary between 0 for pure rotation and 1 for pure elongation. A value of 0.5 corresponds to the case of simple shear.

The volumetric distributions for the parameter λ and shear stress were calculated. For the conveying elements, lower values for both λ and shear stresses can be found in most of the region. Only in the clearances of the intermeshing zone and in the flight clearance both λ and the shear stress have much higher values, indicating that dispersive mixing is more efficient. In order to estimate the contribution of the leakage flows through these clearances to the overall flow, the flow rates of the calender leakage Q_c, the flight leakage Q_f, the side leakage Q_s and the tetrahedron leakage Q_t were calculated. The total leakage $(Q_c+Q_f+Q_s+Q_t)$ can be about 25 % of the total flow rate.

The volumetric distributions of λ and shear stress in the shearing discs region were calculated and compared with those in the conveying elements region for the same rpm and flow rate (Figure 2).

The average values for the parameter λ and shear stress were calculated by weighing the corresponding parameters for each element by the volume of the element itself. Figure 3 presents the average values in conveying elements and shearing discs at various rpms. At different rpms, different values for the axial pressure gradient were specified on the shearing discs in order to match the flow rate of the conveying elements region. The shearing discs are more effective than the conveying elements for dispersive mixing as indicated by the higher overall and average values for both λ and the shear stresses generated.

DYNAMIC STUDY OF DISTRIBUTIVE MIXING IN THE C-CHAMBER

The distributive mixing process was studied by tracking the motion of the minor component particles in the flow field. We assume that the particles can be represented as massless points whose presence does not affect the flow field. Also we neglect the interactions among the particles. Based on the flow field calculated, the locations of the particles can be determined by integrating the velocity vectors with respect to time:

$$\underline{X}(t) = \underline{X}(t_o) + \int_{t=t_o}^{t'=t} \underline{V}(t')dt' \qquad (5)$$

where $\underline{X}(t)$ is the location of a particle at any time t, $\underline{X}(t_o)$ is the location of the same particle at t_o and $\underline{V}(t)$ is the corresponding velocity vector of the particle.

The trajectories of the particles are represented in the series of plots in Figure 4. Initially, 10000 particles were randomly grouped in 10 spherical shaped clusters with the same ratio of particle number to sphere volume. The clusters were randomly placed into the C-chamber. As the screw rotates, the C-chamber moves in the axial direction, and the particles spread away from their initial positions. Figure 4b shows the particle distribution in the C-chamber when the C-chamber moved over one pitch distance in the axial direction (corresponding to one turn of the screw rotation). With the C-chamber advancing in the axial direction, the particles spread gradually throughout the whole volume as shown in Figures 4c&4d.

The goodness of distributive mixing was characterized by the pairwise correlation function. Wong and Manas first used this criterion to study the distributive mixing in a Banbury mixer (13). Distributive mixing was measured by recording the changes in the distance r between pairs of particles. The discrete version of the pairwise correlation function (14) was used for the massless points of particles:

$$f(r) = \frac{2}{N(N-1)} \sum_i \delta(\underline{r_i'} + \underline{r})\delta(\underline{r_i'}) \qquad (6)$$

where $f(r)$ is the coefficient of the correlation function for a separating distance between particles ranging from $r - \Delta r / 2$ to $r + \Delta r / 2$. The summation is performed over all possible but non-repeating pairs of particles; i.e., the number of pairs for a system consisting of N particles is $N(N-1)/2$, which is exactly the inverse of the normalizing factor in $f(r)$ and $\delta(\underline{r})$ is defined as

$$\delta(\underline{r}) = \begin{cases} 1 & \textbf{if a particle is present} \\ 0 & \textbf{if a particle is absent} \end{cases} \qquad (7)$$

For small enough Δr, $f(r)$ can be represented as

$$f(r) = c(r)\Delta r \qquad (8)$$

where $c(r)$ is the coefficient of the probability density function to find a neighbor which is at a distance ranging from $r - \Delta r/2$ to $r + \Delta r/2$. Furthermore, the following identity can be written as

$$\sum_{r=0}^{r=r_{max}} c(r)\Delta r = 1 \qquad (9)$$

where r_{max} is the largest dimension of the system such that $c(r)$ is always zero for $r > r_{max}$; i.e., there is no correlation between points that are separated by distances that are larger than r_{max}.

For a system with perfect distributive mixing, the minor component particles should be randomly distributed throughout the whole mixing region. An ideal distribution was defined as the one which consists of the same number of particles and the particles are randomly distributed throughout the C-chamber. The probability density distribution of the pairwise correlation function for the ideal case and for the case where the C-chamber moved axially one pitch are presented in Figure 5. It should be pointed out that the correlation function depends on the geometry of the flow field (e.g. r_{max} is the longest dimension of the C-chamber and the peak in the curve is due entirely to the presence of the screw root). Also, as mixing time progresses (or as the C-chamber advances axially) the actual correlation function approaches the ideal case. The difference between the actual correlation function and the ideal one was quantified in terms of a parameter ε defined by

$$\varepsilon = \int_{r=0}^{r_{max}} \frac{\left| c(r) - c(r)_{ideal} \right|}{2} dr \qquad (10)$$

where $c(r)$ is the coefficient of the correlation function for our simulations, $c(r)_{ideal}$ is the corresponding coefficient for the ideal case. The normalizing factor 2 was used to make ε range from zero to one.

Figure 6 plots the evolution of ε with the axial distance for various rpms. The parameter ε is large initially and it diminishes gradually with the axial distance or mixing time. The screw rotational speed affects the value of ε only in an initial stage. After about 3 pitches the value of ε remains unchanged, regardless of the processing conditions.

CONCLUSIONS

The 3-D flow patterns in the conveying elements and shearing discs of an intermeshing counterrotating twin screw extruder were obtained by using FIDAP. The simulations were carried out for a power-law model fluid under isothermal conditions. The dispersive mixing efficiency was characterized in terms of shear stresses generated and a parameter λ quantifying the elongational flow components. In the conveying elements region, the various leakage flows through the clearances enhance dispersive mixing. Overall, the shearing discs are more effective than the conveying elements for intensive mixing. The dynamics of distributive mixing in the C-chamber was studied by means of a particle-tracking algorithm. The goodness of distributive mixing was evaluated in terms of a parameter ε measuring the departure from ideality of a pairwise correlation function.

ACKNOWLEDGMENT

The authors would like to acknowledge the financial support of this work by EPIC, and the use of computing services from the Ohio Supercomputer Center.

LITERATURE CITED

1. Dobozcy, Z., Plast. Verarb., **16**, 57 (1965)

2. Kim, W.S., W.W. Skatschkow, and S.D. Jewmenow, Plaste u Kautschuk, **20**, 696 (1973)

3. Janssen, L.P.B.M., L.P.H.R.M. Mulders, and J.M. Smith, Plast. Polym., **43**,93 (1975)

4. Wyman, C.E., Polym. Eng. Sci., **15**, 606 (1975)

5. Konstantinov, V.N, and A.N. Levin, Khim. Mashinostr., **3**, 18 (1962)

6. Konstantinov, V.N, Khim. Neft. Mashinostr., **2**, 21 (1964)

7. Janssen, L.P.B.M., Twin Screw Extrusion, Elsevier, Amsterdam (1978)

8. Speur, J.A., H. mavridis, J. Vlachopoulos, and L.P.B.M. Janssen, Adv. Polym. Tech., **7**, 39 (1987)

9. "FIDAP Package," Fluid Dynamics International, Inc., Evanston, III.

10. Elmendorp, J.J., Polym. Eng. Sci., **26**, 418 (1986)

11. Manas-Zloczower, I. and D.L. Feke, Inter. Polym. Proc., **II**, 185 (1988)

12. Manas-Zloczower, I. and D.L. Feke, Inter. Polym. Proc., **IV**, 3 (1988)

13. Wong, T.H. and I. Manas-Zloczower, Submitted for Publication

14. Witten, T.A. and L.M. Sander, Phys. Rev. Lett., **47**, 1400 (1981)

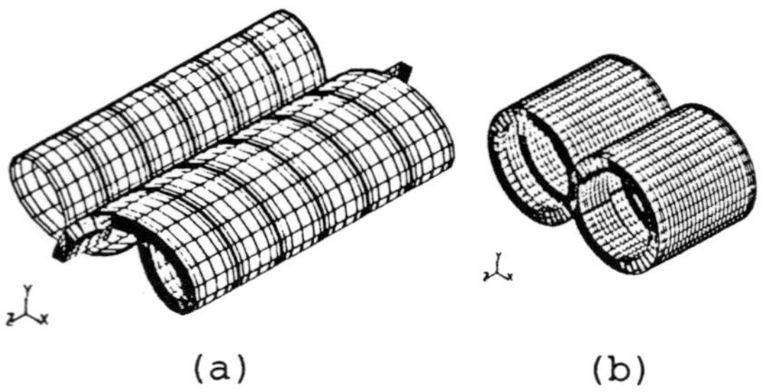

Figure 1. Mesh design for (a) conveying elements region, (b) shearing discs region

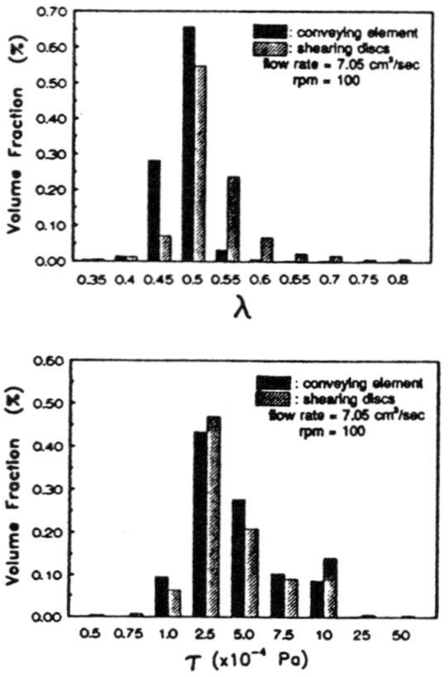

Figure 2. Volumetric distributions of λ and shear stresses in the regions of conveying elements and shearing discs

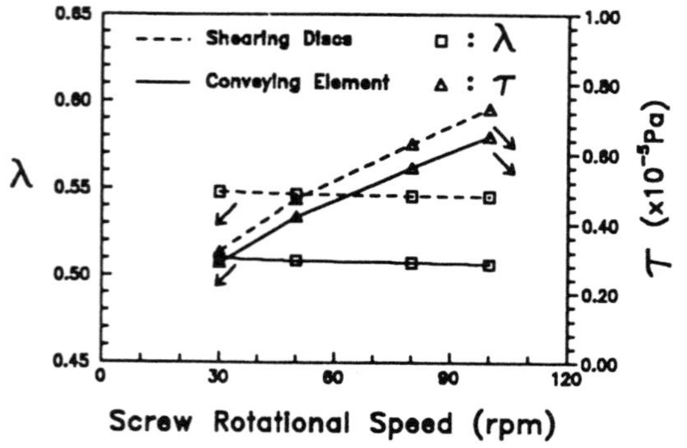

Figure 3. Comparison of the average values for λ and shear stress between the conveying elements and the shearing discs

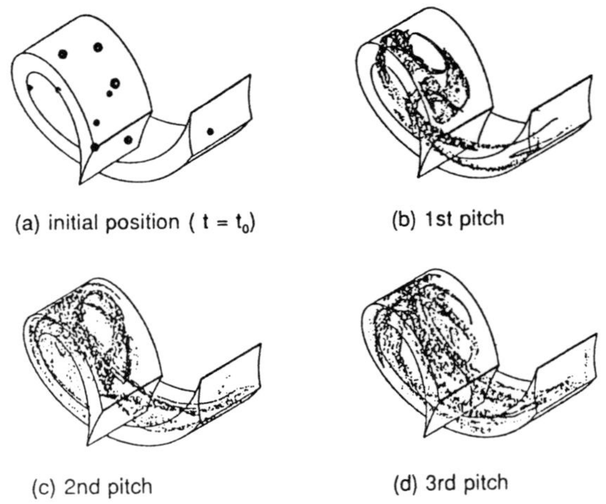

Figure 4. Evolution of particle trajectories in the C-chamber

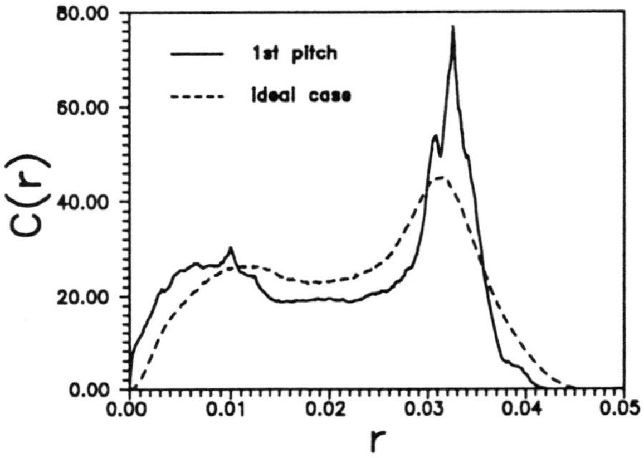

Figure 5. Correlation function

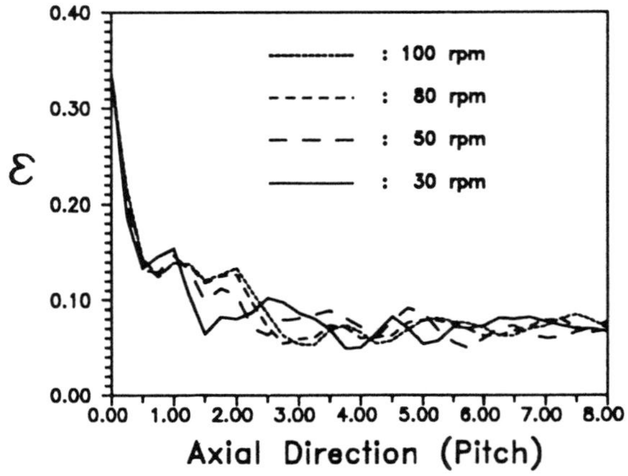

Figure 6. Evolution of ε at various screw rotational speeds

A Moving-Deforming-Mesh Technique for Simulation of Flow in Mixing Tanks

Chin Yuan Perng and Jayathi Y. Murthy

Fluent Inc., Lebanon, NH 03766

This paper describes the development of a moving-deforming mesh approach for the simulation of unsteady flows in mixing tanks. A single grid spans both stationary and moving parts. The grid attached to the impeller moves with it and causes mesh deformation. The numerical procedure employs primitive variables. The discrete equations are obtained by volume integration of the conservation equations over the control volumes in a non-staggered grid system using generalized boundary-fitted coordinates. Spurious pressure modes inherent to non-staggered grid formulations are suppressed by using a contributed momentum approach. The fully implicit-Euler scheme is used for time stepping. Two-and three-dimensional unsteady simulations of laminar and turbulent flows in baffled mixing tanks with a Rushton-type impeller are presented.

INTRODUCTION

Flows in mixing tanks are complex. The unsteady nature of the flows, the complex blade geometries and the relative motion among elements combine to make physical measurements difficult. Numerical approaches hold great promise but must be able to deal with geometric complexity and the interaction of the rotating turbine blades and the stationary baffles along the tank periphery. The presence of the baffles makes the flow inherently unsteady since the computational domain cannot be immobilized by using a frame of reference rotating with the blades. Consequently, many approaches in the literature simplify the problem by either removing the baffles altogether or by representing their influence in some axi-symmetric way. For example, Hutchings et al. [1] modeled the effect of the tank baffles by using an axisymmetric sink of circumferential momentum at the tank walls to obtain steady state solutions in a reference frame rotating with the impeller. However, this approach is limited to simple tank geometries with negligible impeller-baffle interactions.

Both sliding-mesh and deforming-mesh techniques may be taken in dealing with internal moving boundaries. In the sliding-mesh approach, two separate grids are used; one moves with the rotating part and the other fixed to the tank. The two grids move relative to each other, slipping along a common interface. In the deforming-mesh method, single grid is used to span both stationary and moving parts and allows the grid to deform with the moving part. The former approach was used by Rai [2]. Rai employed patched and/or overlaid grids to perform both Euler and Navier-Stokes simulations for rotor-stator interactions in turbomachinery. Excellent results for 2-D calculations were reported. However, a great deal of bookkeeping is required to keep track of the addresses of adjacent neighbours at every time step. So far, little success has been reported in 3-D calculations.

In this paper, we use a single mesh for both stationary and moving parts. The grid attached to the rotor moves with it and causes the interfacial mesh to deform. When mesh deformation becomes acute, the grid is regenerated locally and the flow variables are transferred to the new mesh in a conservative fashion. The grid motion is accounted for by transforming the time derivative in Eulerian conservation equations into a Lagrangian form [3, 4]. In both papers, only global domain boundaries were moved in prescribed ways. The procedures we describe here are especially tailored for internal moving boundaries.

In the sections that follow, we present the governing equations for use with moving meshes. We describe the techniques for mesh deformation, regeneration and variable interpolation. We then use the techniques described to compute both laminar and turbulent flows in sample mixing tanks.

FORMULATION

The integral form of equations governing the incompressible Newtonian fluid in a time-varying control-volume $V(t)$ enclosed by moving surface $\mathbf{S}(t)$ can be written as:

$$\frac{d}{dt}\int_{V(t)} dV = \oint_{\mathbf{S}(t)} \mathbf{v}\cdot d\mathbf{S} \tag{1}$$

$$\frac{d}{dt}\int_{V(t)} \rho dV + \oint_{\mathbf{S}(t)} \rho(\mathbf{u}-\mathbf{v})\cdot d\mathbf{S} = 0 \tag{2}$$

$$\frac{d}{dt}\int_{V(t)} \rho\mathbf{u} dV + \oint_{\mathbf{S}(t)} \left[\rho(\mathbf{u}-\mathbf{v})\mathbf{u} - \overline{\mathbf{T}} - \overline{\mathbf{T}}_{\mathbf{t}}\right]\cdot d\mathbf{S} = \int_{V(t)} \mathbf{s}_u dV \tag{3}$$

$$\frac{d}{dt}\int_{V(t)} \rho\phi dV + \oint_{\mathbf{S}(t)} \left[\rho(\mathbf{u}-\mathbf{v})\phi - \mathbf{q} - \mathbf{q}_{\mathbf{t}}\right]\cdot d\mathbf{S}$$

$$= \int_{V(t)} s_\phi dV \quad (4)$$

where ρ is the fluid density, $\mathbf{u}$ is the flow velocity vector, and $\mathbf{v}$ is the control volume surface velocity vector arising from the motion of moving mesh. $\overline{\mathbf{T}}$ is the molecular stress tensor for Newtonian fluids and $\overline{\mathbf{T_t}}$ is the Reynolds stress tensor $(-\rho\overline{\mathbf{u}'\mathbf{u}'})$. $\mathbf{q}$ and $\mathbf{q_t}$ are the molecular diffusion and the turbulent transport fluxes of scalar quantity ϕ. The $k-\epsilon$ two equation model together with Boussinesq's gradient transport concept are used for turbulence closure [5]. $\mathbf{s}_u$ and s_ϕ are the source terms for $\mathbf{u}$ and ϕ equations, respectively. Note that d/dt in all above expressions is the total derivative. It represents the time rate of change of a variable "seen" by an observer riding on the moving control volume.

An important component of the calculation procedure is the inclusion of Eqn. (1) for space conservation. Eqn. (1) states that the time rate of change of a cell volume is equal to the total volume swept by its enclosed surface $\mathbf{S}$ [3, 4]. A theoretically equivalent view on Eqn. (1) is to follow the material coordinates of the individual cell and make use of the Jacobian of transformation, J. We thus obtain an equivalent expression $dJ/dt = J\nabla\cdot\mathbf{v}$ [6]. This expression is known as "the moving mesh metric identity" which links the general Eulerian conservation equations to Eqns. (2) to (4). We use Eqn. (1) to numerically determine the updated value of the cell volume (i.e., the Jacobian J) at each time step to avoid possible creation of spurious volume sources and/or sinks which may corrupt the entire solution [7].

NUMERICAL METHOD

The numerical procedure solves for Cartesian velocity components with boundary-fitted coordinates. The discrete equations are obtained by volume integration of the conservation equations (Eqns. 1 to 4) over each control volume. A second-order central differencing scheme is used for the diffusion terms. Both the power law [8] and QUICK [9] are employed for the convection terms. Time marching is accomplished through the fully implicit Euler scheme. The basic procedure is similar to that of Patankar [8] and is not elaborated further.

The formulation uses primitive variables on a non-staggered grid system. A procedure similar to that of Rhie and Chow [10] is used to suppress spurious pressure modes, with the corrections specified by Majumdar [11] to ensure that the final solution is independent of under-relaxation factors and time steps. The SIMPLE algorithm [8] is used for pressure-velocity coupling. Since these procedures are described in the literature, we do not elaborate them further. In the sections below, we consider specific issues related to the moving mesh problem.

Strategy for Mesh Moving and Deforming

Figure 1 displays a sequence of deforming and moving meshes. At time $t = 0$, the calculation domain is meshed with a single grid as shown in Figure 1a. A rotationally cyclic boundary condition is imposed on the right and left boundaries. The blades move in a counter-clockwise fashion and carry the attached mesh with them. In this example, we choose to deform only a single row of cells, as shown in Figure 1b, though it is possible to transmit deformation to the entire grid. When the deformation of these interface cells reaches a preset limit, Figure 1b, we "snap" the grid back by one cell, as in Figure 1c. Since the right and left boundaries are cyclic, the rightmost grid-line together with its corresponding cells are "snapped back" to the leftmost grid line. At this point, all the cells and grid nodes involved in moving and snapping operations are re-numbered and the procedure continues. Thus, a continuous grid motion is achieved by the sequential combination of mesh motion, deformation, snapping, and grid re-numbering.

There are several advantages to the "moving and deforming" strategy. The most important one is that it does not require a sophisticated data structure and elaborate bookkeeping to track the correspondence between the stationary and moving parts of the grid. Snapping, renumbering, and interpolation take place locally and are relatively inexpensive.

There are shortcomings associated with this strategy. The major one is that the interfacial cells are allowed to skew, and solution accuracy may degenerate if grids are not snapped often. However, this disadvantage can be overcome by grid smoothing and by choosing the line of deformation in regions where the variations of flow variables are deemed small.

Interface Interpolation

In a non-staggered grid system, both the velocity vector and the flow variable ϕ sit at the center of the control volume, while the mass fluxes locate at control volume faces. All these quantities must be interpolated after grid snapping. Flow variables are required at the snapped locations to serve as initial conditions for subsequent time-stepping. Mass fluxes are required only as initial guesses for the non-staggered procedure. In the present work, a mass-averaged interpolation is used for the interpolation of velocity components and flow variables. The face mass fluxes are obtained through mass conservation over the appropriate cell region.

APPLICATIONS

Two-Dimensional Mixing Tank with Baffle

The geometry and computational mesh for a two-dimensional stirred tank are shown in Figure 2a. The tank has four equally spaced baffles; the impeller has four blades. Consequently, only one 90-degree sector with one wall baffle and one impeller blade need be considered, with rotationally cyclic boundary conditions applied on the left and right boundaries. The ratio of the inner to outer radius is 0.18; the dimensionless blade-tip radius (based on the outer radius) is 0.51; the dimensionless thickness of the baffle is 0.16. The flow is assumed to be laminar and initially quiescent, with the blade aligned with the baffle. The rotor is impulsively started and sets up a flow field. A 43 x 42 nonuniform grid is used to discretize the domain. Figures 2b to 2d depict a sequence of flow fields as the blade moves on. Tip vortices are formed at start-up and are shed as the rotation proceeds. In this example, the wall baffles have little effect on the flow field. The vortices die down before they can be transported to the baffle region.

Three-Dimensional Mixing Tank with Baffle

The aspect ratio of the tank (height/outer radius) is 2.5 and contains a Rushton impeller with dimensionless tip radius (based on the tank radius) of 0.325. There are four equally spaced baffles and the impeller also has four equally-spaced blades. Therefore, we consider a 90-degree sector with rotationally cyclic boundary conditions imposed on the two sector faces. A no-slip condition is imposed on all walls except the top one; the top wall is considered to be a fluid free-surface and modeled by requiring that there be no normal velocity and a zero gradient in the tangential velocity. The rotational Reynolds number, based on the blade tip velocity and the tank diameter, is 2×10^4. The computational domain is discretized using a $21 \times 19 \times 24$ grid in the circumferential, radial and axial directions, respectively. The moving part of the three-dimensional grid consists of a cylindrical region with a $21 \times 9 \times 6$ grid which encloses the impeller region. This inner grid moves with the speed of the impeller.

The tank flow field predicted by the moving mesh simulation are illustrated by both velocity vectors and particle tracks. Figure 3a displays three-dimensional velocity vectors on a horizontal plane right above the blades, and Figure 3b shows a similar plot right below the blades. As expected, the flow is thrown outwards by the blades and sucked in near the tank axis, causing helical vortex motion in the circumferential direction.

Figure 3c displays the circumferential vortex motion through particle trajectories in mean flow velocity field at a given time instant. The spiral vortex motion induced by the blade is clearly visible. This pattern is typical of three-dimensional mixing tanks [1] with the primary displacement of fluid being axial.

CONCLUSIONS

A moving-mesh technique for the simulation of flow fields in mixing tanks has been presented. The numerical method is validated against calculations in a rotating frame and subsequently applied to unsteady laminar and turbulent flows in two- and three-dimensional mixing tanks. Characteristic flow features are correctly captured.

ACKNOWLEDGMENTS

We thank Fluent Inc. of Lebanon, New Hampshire for the use of the FLUENT computer code as the basis of the work presented in the paper.

References

[1] Hutchings, B.J., B.R. Patel, and R.J. Weetman, "Computational of Flow Fields in Mixing Tanks with Experimental Verification," ASME Annual Winter Meeting, San Francisco, CA (1989).

[2] Rai, M.M., "Navier-Stokes Simulations of Rotor--Stator Interaction Using Patched and Overlaid Grids," *AIAA-85-1519* (1985).

[3] Demirdzic, I. and M. Peric, *Int. J. Num. Methods in Fluids*, **10**, 771 (1990).

[4] Rosenfeld, M. and D. Kwak, *Int. J. for Num. Methods in Fluids*, **13**, 1311 (1991).

[5] Rodi, W., "Turbulence Models and Their Application in Hydraulics – A State of the Art Review," the *IAHR-Section on Fundamentals of Division II: Experimental and Mathematical Fluid Mechanics*, (The Netherlands) (1980).

[6] Aris, R., "Vectors, Tensors, and the Basic Equations of Fluid Mechanics," Dover 0-486-66110-5 (1989).

[7] Thompson, J.F. and C.W. Warsi, "Numerical Grid Generation; Foundations and Applications," Elsevier Science (1985).

[8] Patankar, S.V., "Numerical Heat Transfer and Fluid Flow," McGraw Hill New York (1980).

[9] Leonard, B.P., *Compu. Methods in App. Mech. and Eng.*, **19**, 59 (1979).

[10] Rhie, C.M. and W.L. Chow, *AIAA J.*, **21**, 11, 1525 (1983).

[11] Majumdar, S., *Numerical Heat Transfer*, **13**, 125, (1988).

NOMENCLATURE

J	the jacobian of transformation
$\mathbf{q}$	molecular diffusion flux vector
$\mathbf{q_t}$	turbulent transport flux vector
$\mathbf{S}$	cell face vector
$\mathbf{s_u}$	source term for $\mathbf{u}$ equation
s_ϕ	source term for ϕ equation
$\overline{\overline{\mathbf{T}}}$	molecular stress tensor
$\overline{\overline{\mathbf{T}}}_\mathbf{t}$	Reynolds stress tensor
t	time
$\mathbf{u}$	flow velocity vector
$\mathbf{v}$	cell face velocity vector
V	cell volume
ρ	fluid density

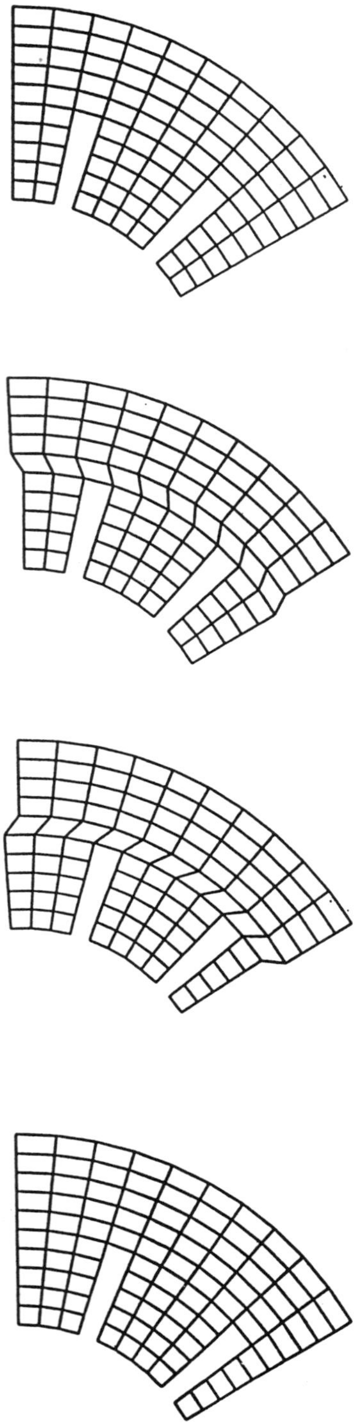

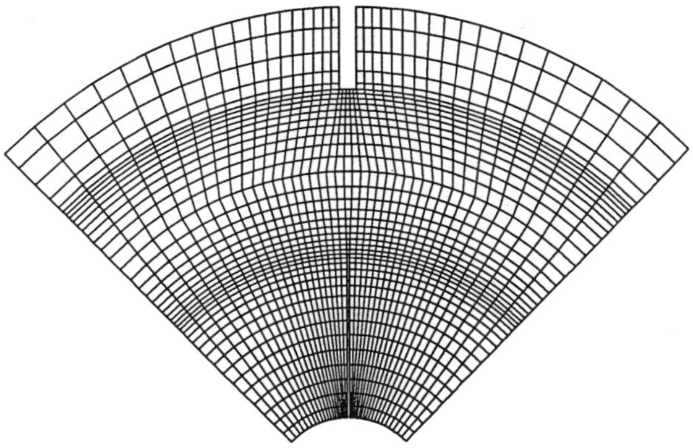

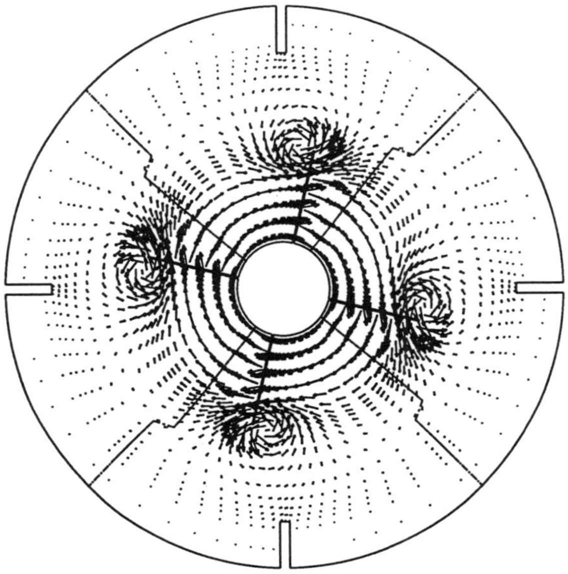

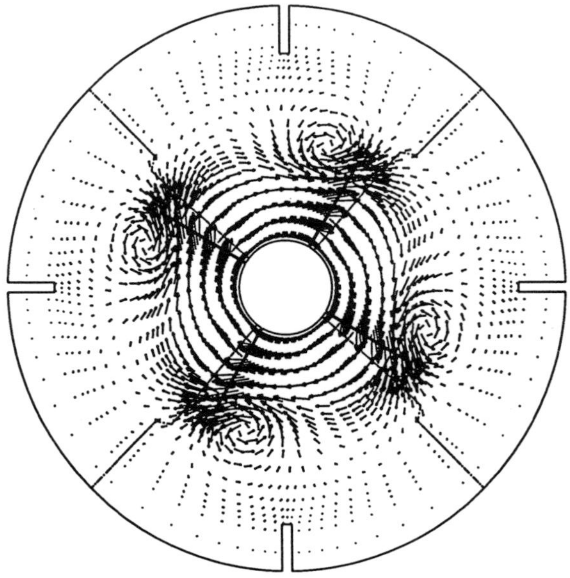

Figure 1. Grid motion through deformation and snapping: a) initial grid, b) grid moves and deforms, c) the right boundary and cells are snapped to the left, d) grid advances one cell distance.

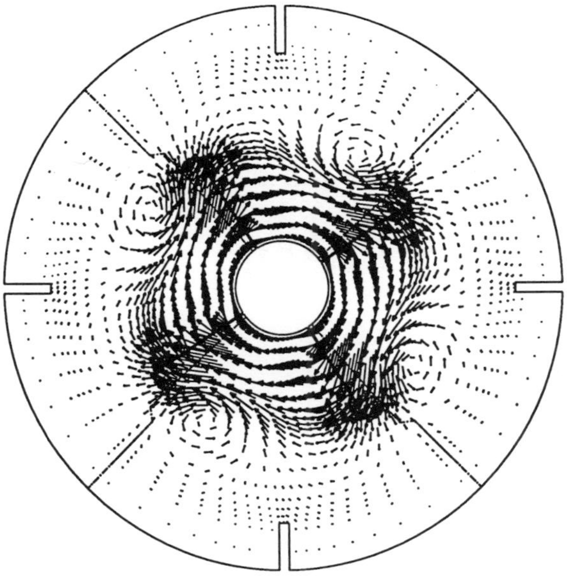

Figure 2. Geometry and developments (from "b" to "d") of tip vortices in a two-dimensional baffled mixing tank at $Re = 1000$.

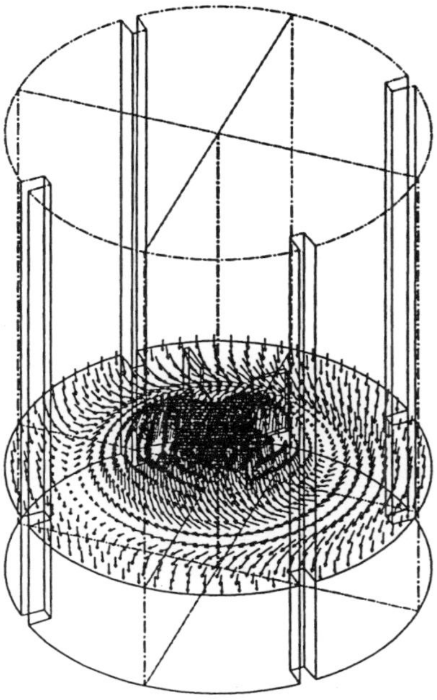

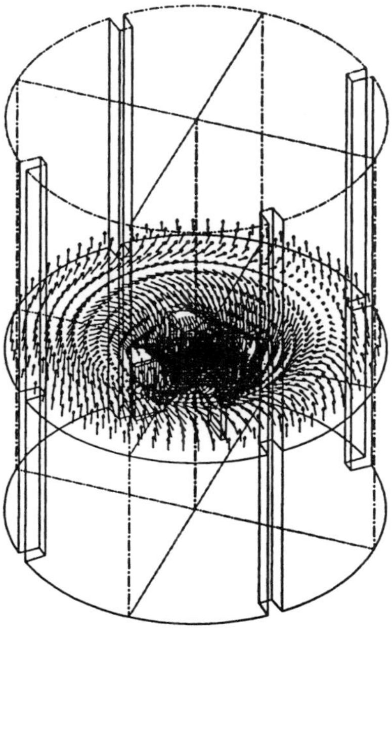

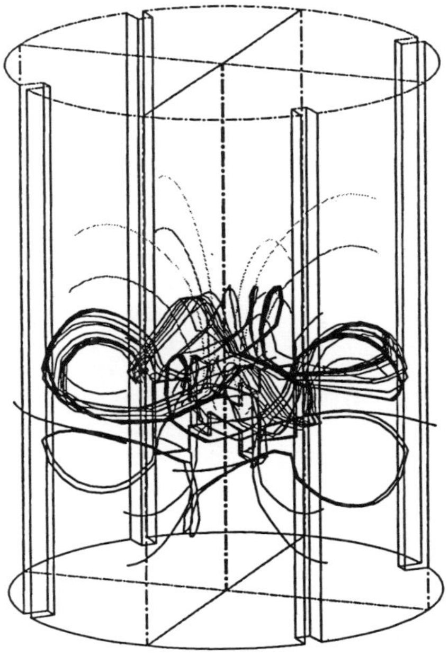

Figure 3. Flow field in a three-dimensional baffled mixing tank at $Re = 2 \times 10^4$: a) velocity vectors above the impeller, b) below the impeller, c) particle trajectories of the mean flow field.

The Numerical Simulation of Chaotic Mixing Processes

E. P. L. Roberts

EA Technology, Capenhurst, Chester, CH1 6ES, U.K.

A range of Lagrangian techniques are presented for studying the mixing behavior of chaotic flows. The mixing mechanisms were observed by following the motion of regions of fluid modelled as arrays of passive marker particles. The dominant mixing mechanism was observed to be a successive stretching and folding process. Techniques for quantifying the mixing rate in chaotic flows were tested and an approach based on the stretch rate of line elements was found to be most suitable. This gives a volume averaged non-diffusional mixing measure which is independent of the numerical parameters. The variation in the local stretch rate can be used to follow the mixing distribution within chaotic flows using stretch rate maps.

Mixing processes have traditionally been modelled by following the concentration distribution for comparison with experimental observations of (for example) mixing times. This approach is limited in that it is numerically difficult to model concentration fields when the diffusion coefficient is low (e.g. with many liquid flows). Furthermore there is a requirement for an arbitrary initial concentration distribution and this makes the comparison of mixing rates in different geometries rather difficult. Little advancement has been made by this route in the understanding of fluid mechanical mixing mechanisms and it seems likely that for many situations the experimentally observed mixing rate will be controlled by the non-diffusional mixing processes. It is therefore desirable to develop new techniques that separate dynamical non-diffusional fluid mixing from the diffusional processes. This type of approach would facilitate the scale up of mixing devices, and allow the comparison of different mixing flows without the need to consider diffusional processes.

Although some of the early work on mixing processes recognised the need to observe Lagrangian processes e.g. Batchelor [1], Welander [2], detailed studies of non-diffusional mixing in even simple flows has only recently been achieved. Developments in the field of dynamical systems have led to a new approach to the understanding and study of the kinematics of fluid mixing. With the publication of the book by Ottino [3] a mathematical framework for the analysis of mixing systems has been put forward. This approach uses Lagrangian techniques to characterise the mixing process by observing the stretch rates of fluid lines or surfaces. If one considers a system of two segregated fluids then, in the absence of diffusion, the only form of mixing that can occur is the stretching of the interface between the two fluids. In the terms of Dankwerts [4] paper this is equivalent to the reduction of the scale of segregation, and therefore a measure of the kinematic mixing rate.

Aref [5] studied the advection behaviour of some very simple two dimensional unsteady flows. He observed the motion of a region of fluid in a prescribed flow field (using an array of marker particles) and found that under some conditions the region of fluid became rapidly mixed with the surrounding fluid. This type of mixing behaviour has been termed chaotic mixing. From a process engineering viewpoint there is nothing particularly special about a chaotic flow, indeed the flow behaviour in any efficient process mixing device will, by its very nature, be chaotic (e.g. turbulent flow). Aref [5] clearly identified the importance of using a Lagrangian approach to study the mixing behaviour in this type of flow and has illustrated an important difference between steady and unsteady flow. From a Lagrangian viewpoint elements of fluid are constrained to follow streamlines in a steady two dimensional flow and Eulerian streamline plots can clearly be used to illustrate the Lagrangian (mixing) behaviour. For unsteady flows this constraint is removed and it is necessary to observe the Lagrangian behaviour as well as the Eulerian flow field in order to illustrate the mixing behaviour.

The objectives of the current paper are twofold. The first objective is to develop numerical flow visualisation techniques based on the Lagrangian approach which are appropriate for the study of mixing in process engineering flows. This will help to improve the chemical engineers understanding of mixing mechanisms in process devices, thus aiding the development of new geometries and the improvement of existing designs. The second objective is to develop appropriate techniques for quantifying the kinematic mixing rate based on the Lagrangian stretch rate techniques developed by Ottino [3]. The development of a kinematic mixing measure (independent of diffusion processes) will help the chemical engineer with scale up processes, allow the comparison of different mixing geometries and the optimisation of existing designs.

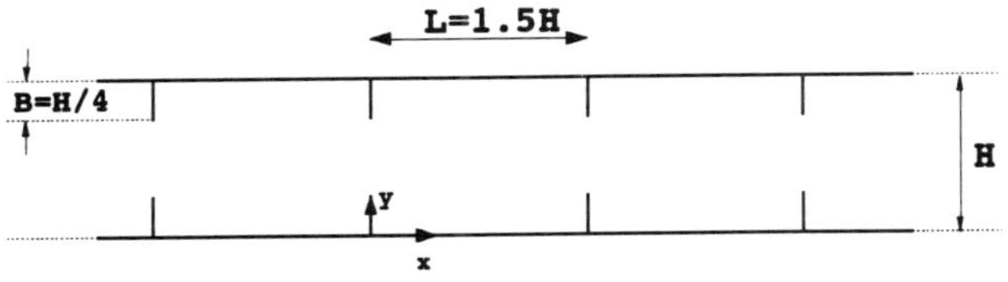

Figure 1. The baffled channel geometry.

The geometry chosen for this study is a two dimensional periodic baffled channel, as presented by Howes *et al.* [6] and illustrated in figure 1. The flow of an incompressible Newtonian fluid has been modelled within this geometry for a range of flow conditions. Two classes of flow have been studied:

(a) constant volumetric flow, characterised by a Reynolds number based on cross sectional mean velocity (U) and channel width (H)

(b) oscillatory flow with fully reversing sinusoidal volumetric flow, characterised by a Reynolds number based on peak cross sectional mean velocity (U) and channel width, and a Strouhal number - $St = \Omega H/U$ (where Ω is the frequency of oscillation in Hz).

The geometry is typical of many heat transfer surfaces with baffles giving increased surface area and turbulence enhancement. Oscillatory flows in this geometry are known to give excellent radial mixing and surface transfer properties at moderate Reynolds numbers (Dickens *et al.* [7], Mackley *et al.* [8]) and are amenable to numerical simulation (Howes *et al.* [6]).

The numerical simulation is a finite difference model based on vorticity and stream function and the flow is assumed to be spatially periodic, with the flow within each cell identical. A finite difference grid with 42 x 65 grid points has been used with a time step of $tU/H = 0.005$. Details of the numerical simulation are described by Howes *et al.* [6] and Roberts [9].

FLOW PATTERNS

For constant volumetric flows the flow is symmetric and steady at low Reynolds numbers with recirculating regions downstream of each baffle. The length of the recirculating region grows approximately linearly with Reynolds number until, at N_{Re}~80, the reattachment point moves on to the downstream baffle and the recirculating region occupies most of the inter-baffle space. At a Reynolds number of ~100 the flow becomes asymmetric and unsteady with eddies shed alternately from top and bottom baffles. In this regime the flow is periodic. With increasing Reynolds number the flow becomes increasingly asymmetric but remains periodic up to a Reynolds number of 400.

Oscillatory flows show vortex formation, growth and ejection downstream of the baffles for each half cycle of the oscillation. A pair of eddies is thus formed during each half cycle of oscillation and the level of vortex interaction increases with increasing Reynolds number. Similar flow patterns have been described by Sobey [10] for oscillatory flow in wavy walled channels and Brunold *et al.* [11] for oscillatory flow in periodically baffled tubes. At a Reynolds number of ~100 the flow becomes asymmetric with a dominating eddy in each cell. This dominating eddy forces the mainstream flow to sweep alternately across the top and bottom walls during each oscillation cycle.

For unsteady flows a sequence of instantaneous streamline plots (or Eulerian velocity fields) can only provide highly qualitative information on the mixing behaviour as the scale of motion between plots is not apparent. Lagrangian flow visualisation must be used in order to illustrate the mixing mechanisms.

NUMERICAL VISUALISATION

In order to study the kinematic mixing mechanisms in the absence of diffusion it is necessary to follow the motion and interaction of regions of fluids. Two approaches have been developed.

The first of these techniques involves modelling a region of fluid as an array of passive marker particles which are advected by the flow field. The location of each particle is obtained by integrating the local velocity field over each time step. Following Aref [5], a localised 'blob' of particles can be introduced into the flow and the motion of the particles will illustrate how a localised region will interact with the surrounding fluid.

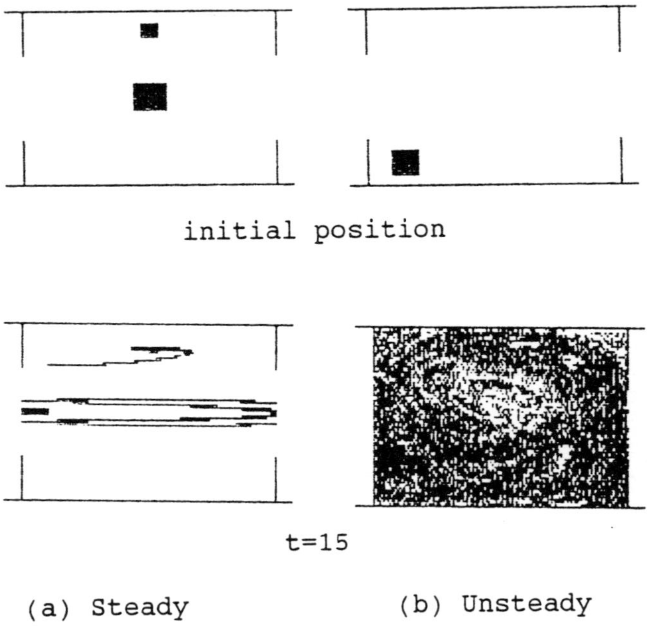

Figure 2. Advection of arrays of particles for constant volumetric flow.

Figure 2 shows the behaviour of blobs of particles for constant volumetric flow for both the steady (figure 2a) and unsteady (figure 2b) regimes. In order to give a compact image the position of the particles is plotted in a single cell so that each particle at position (x+nL, y) where n is an integer, is plotted at position (x,y). Due to the periodicity of the geometry this is equivalent to having identical starting conditions in all adjacent cells. For the steady flow the particles are constrained to follow the streamline on which they lie, and a blob in the recirculating region between the baffles is trapped in the inter-baffle region. The blob of particles in the centre of the channel is constrained by its bounding streamlines and is simply stretched as it is swept downstream. For the unsteady flow there is a dramatic change in the advection behaviour and the blob of particles has become mixed over a wide area, including regions close to the walls and in to the corners of the baffles. These observations are similar to those of

Aref [5] and show the dramatic difference in the advection behaviour that can occur between steady and unsteady flows. The strong mixing behaviour has been termed chaotic advection and is known to lead to enhanced transfer properties as well as efficient mixing.

The 'blob' approach can give qualitative information on the mixing behaviour but is limited in that it requires an arbitrary starting condition (blob location) and cannot illustrate the dominant mixing mechanisms and structures within a chaotic flow. In order to improve the visualisation of the mixing mechanisms an array of particles has been place across an entire cell and different regions of fluid identified using colour graphics (Howes *et al.* [6]).

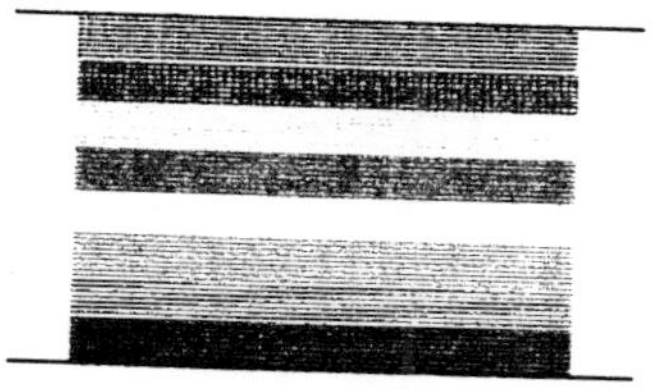

(a) Initial positition

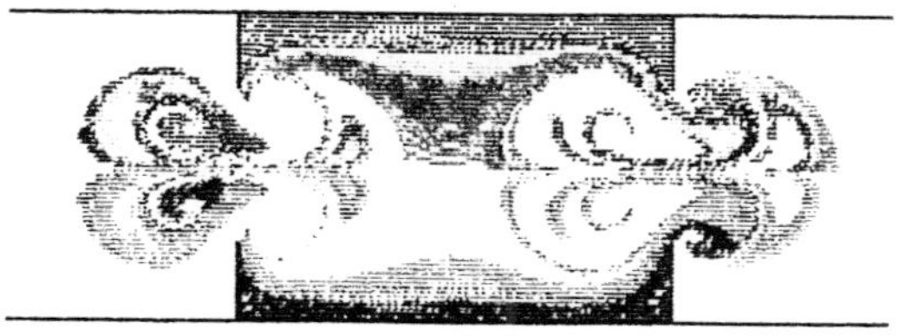

(b) t=1.0

Figure 3. Advection of layers of particles for oscillatory flow with N_{Re}=80, St=1.0.

Figure 3 shows the advection of layers of fluid for an oscillatory flow with N_{Re}=80. The eddies formed during each half cycle of oscillation roll up the layers of fluid and these are then stretched as they are swept between the baffles. This successive folding and stretching mechanism was observed for all of the chaotic flows studied. Thus for the oscillatory flows strong mixing regions are observed in the eddy shedding regions up and downstream of each baffle. In the centre of the cell where the fluid does not advect between the baffles the mixing rate can be seen to be low (figure 3). The coloured layer approach can thus be used to observe mixing mechanisms and structures in chaotic flows (Howes *et al.* [6]).

In the absence of diffusion a region of fluid can essentially be defined by its bounding interface. Thus the motion of a region of fluid can be followed by modelling the motion of interfaces. A technique has been developed where an fluid line is modelled as a chain of particles. If two particles become too far apart then extra particles are interpolated between them. The technique was observed to model the motion of lines well, with the detailed shape of the interface retained. However, for chaotic flows the line stretches exponentially and the number of particles required to model the interface rapidly becomes unwieldy. By modelling lines of limited length (and removing particles from the ends of the line when it became too long) it was possible to observe consistent structures within the chaotic flows. These structures were found to correspond with unstable manifolds within the flow, located using conventional techniques (Roberts, [9]). This is consistent with the paper by Rom-Kedar *et al.* [12] who showed that "...... the unstable manifold is the dominant organising structure ....." in many chaotic flows. The line modelling technique was also found to be useful for establishing the detailed structure of the unstable manifolds themselves.

CHARACTERISATION OF MIXING RATES

In order to quantify the kinematic mixing rates within the chaotic flows studied four different approaches were developed.

Particle Concentration Method

A coarse grid is laid out over the cell and the concentration of particles in each grid square is obtained by particle counting. A blob of particles is advected in the flow field and an intensity of segregation is defined (based on the Dankwerts, [4], approach to the concentration field). By careful consideration of the mechanisms involved a mixing rate parameter can be defined (based on the rate of decrease of the intensity of segregation) which is in principal insensitive to the numerical parameters (Roberts [9]). This mixing rate has been found to be numerically stable but the rate obtained is sensitive to the arbitrary starting condition and shows a high degree of scatter. The use of an arbitrary starting condition makes the comparison of different geometries impractical. The technique could not reliably give a volume averaged measure but is dependent on the initial location of the blob and the distribution of mixing rate.

Line Stretch Method

Given that the non-diffusional mixing of two immiscible liquids is governed by the stretch rate of the interface approach has been developed based on the stretch rate of fluid lines. The method uses the line modelling techniques described above in the section on numerical flow visualisation. By monitoring the stretch rate of an array of 14 fluid lines of limitted length it was possible to obtain a time averaged mixing rate. With appropriate choice of parameters the technique could be made numerically stable. The technique required a significantly larger CPU time to obtain a quantified mixing rate when compared with the other techniques tested, and presented some practical difficulties. The time averaged stretch rate of the lines would occasionally fall away instead of reaching a steady state. This was due to lines becoming trapped in low mixing (low stretch rate) regions (e.g. close to the walls).

Local Stretch Rate Method

This approach is similar to the efficiency of mixing technique presented by Ottino [3]. The time averaged stretch rate of an array of infinitesimal line elements which are advected with the flow is obtained from the local velocity gradients around each line element. By laying an array of elements across a whole cell it was possible to obtain a volume averaged measure of the mixing rate within the cell. The technique was insensitive to all numerical parameters and used relatively little CPU time. The volume and time averaged exponential stretch rate for an array of ~36000 line elements was observed to rapidly approach an asymptotic value, which could be used as a quantified mixing measure.

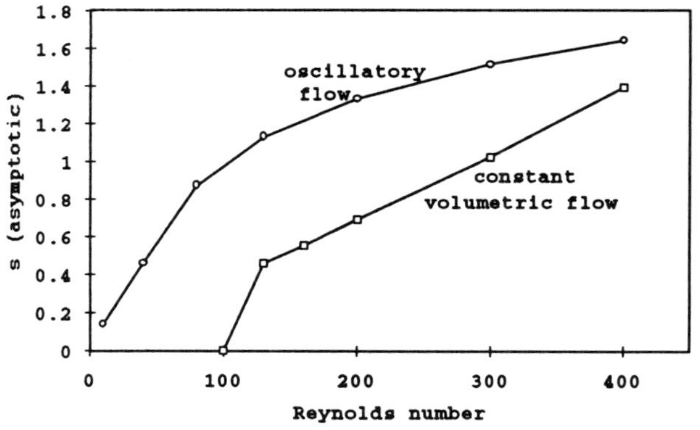

Figure 4. Asymptotic stretch rate versus Reynolds number.

Figure 4 shows the observed asymptotic stretch rate (s) as a function of Reynolds number for both constant volumetric flow and oscillatory flow with St=1.0. For constant volumetric flows the asymptotic stretch rate is zero for steady flow when the advection behaviour is not chaotic. When the flow becomes unsteady there is a step change in the stretch rate observed from s=0 to s~0.4. At higher N_{Re} the stretch rate increases approximately linearly with Reynolds number. For oscillatory flows at low Reynolds the stretch rate increases rapidly as the eddy mixing activity increases. For Reynolds numbers above 150 the stretch rate shows little further increase. Note that as exponential stretch rate has dimensions of 1/time and has been made dimensionless with U/H a constant value of s represents a linear increase in dimensional stretch rate with Reynolds number.

Particle Separation Method

This technique is essentially a numerically simpler version of the line element method, requiring no special modelling techniques. By modelling the motion of an array of particles over the whole cell it is possible to estimate the local stretch rate distribution by observing the separation rate of initially adjacent particles. Over relatively short time scales ($tU/H < 2.0$) the volume and time averaged stretch rate obtained was found to be in good agreement with the rate obtained using the line element technique. Provided the time averaged stretch rate has reached an asymptotic value over this time scale it is possible to obtain a reliable mixing measure using simple particle advection modelling techniques.

Details of all of these techniques and the results obtained are described by Roberts [9].

STRETCH RATE MAPS

Although a single parameter mixing measure is extremely useful, in many cases it will also be important to consider the distribution of mixing. Using the local stretch rate approach it is a simple matter to plot the distribution of stretch rates observed throughout the flow.

Figure 5 shows the stretch rate map for oscillatory flow at N_{Re}=80 (figure 5a) and at N_{Re}=400 (figure 5b) with St=1.0 and for an advection time of t=1.0. At each location throughout the cell the local stretch rate is calculated from the average for two perpendicular elements initially located at this point. The stretch rate map shows the rich mixing structure within the flow and gives a clear depiction of the distribution of mixing within the geometry. Any low mixing regions can be identified by low stretch rate regions. For oscillatory flow the eddy mixing regions extend into the cell so that at a moderate Reynolds number (figure 5a, N_{Re}=80) the stretch rate is high throughout most of the cell. Low stretch rate regions are apparent in the corners of the cell, close to the walls and in the centre of the cell where the fluid is not swept between the baffles. The successive stretching and folding mechanism observed in the eddy shedding regions clearly lead to high stretch rates, efficient and well distributed mixing. At higher Reynolds numbers (figure 5b) where the flow has become asymmetric the stretch rates are high close to the walls (suggesting high wall transport rates) and relatively low in the centre of the cell where the fluid is simply rolling around within a dominating eddy.

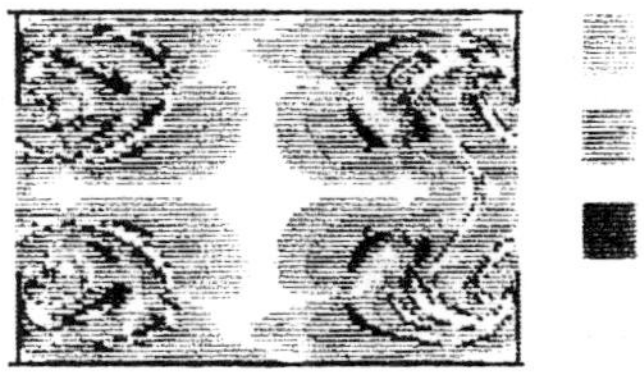

(a) N_{Re} = 80

(b) N_{Re}=400

Figure 5. Stretch rate maps for oscillatory flow, St=1.0.

Clearly stretch rate maps can be used to provide useful information on the distribution of mixing and the mixing structure within a given flow. Other parameters such as the variance of the stretch rate can be used to quantify how evenly the local stretch rate (mixing) is distributed throughout the flow (Roberts, [9]).

DISCUSSION AND CONCLUSIONS

A range of Lagrangian techniques have been developed for numerical visualisation and quantification of kinematic mixing behaviour. For constant volumetric flow within baffled channels the early transition to unsteadiness caused by the presence of the baffles leads to chaotic mixing. Fluid is no longer trapped in recirculating regions and the presence of the baffles will lead to increased wall transport rates (heat and mass transfer), increased mixing and reduced dispersion. The dominant mechanism for mixing was observed to be a successive stretching and folding process leading to rapid stretching and interaction of layers of fluid. Oscillatory flows were observed to give excellent mixing properties as the successive eddy formation and shedding leads to rapid stretching and folding.

The Lagrangian techniques developed can clearly be used to illustrate the mechanisms and structures within mixing flows. This will allow the design and development of novel mixer designs as the engineer's understanding of mixing behaviour is improved. A quantified measure of non-diffusional mixing rates will allow process engineers to directly compare the mixing performance of different geometries and to optimise existing designs. It should for example be possible to optimise the baffle spacing to provide the most rapid mixing for oscillatory flows in baffled channels. As the diffusion process is decoupled from the mixing measure, the difficulties associated with scaling up mixing processes should become simpler. Stretch rate maps can also be used to establish the distribution of mixing in a given geometry and to locate any stagnant or low mixing regions which may not be apparent from Eulerian observations.

All of the Lagrangian techniques developed could easily be included with existing computational fluid dynamic (CFD) packages, many of which already have the facility for modelling particle advection. The CPU time associated with the computations are relatively low and provided the concepts can be easily understood these new techniques will provide a useful tool for chemical engineers using CFD packages.

ACKNOWLEDGEMENTS

The author is grateful to Dr. Malcolm Mackley for many fruitful discussions on oscillatory flow and mixing and to Dr. Tony Howes for allowing the use and adaptation of his flow modelling codes. The work was financed by BP Research (Sunbury) and by the Science and Engineering Research Council.

LITERATURE CITED:

[1] Batchelor, G.K., *Proc.R. Soc. Lond.*, **A213**, 349, 1952

[2] Welander, P., *Tellus*, **7**, 141-156, 1955

[3] Ottino, J.M., "The Kinematics of Mixing", Cambridge Univ. Press, 1989

[4] Dankwerts, P.V., *App. Sci. Res.*, **3**, 279-296, 1952

[5] Aref, H., *J. Fluid Mech.*, **143**, 1-22, 1984

[6] Howes, T., Mackley, M.R. and Roberts, E.P.L., *Chem. Engng. Sci.*, **46**, 1669-1677, 1991

[7] Dickens, A.W., Mackley, M.R. and Williams, H.R., *Chem. Engng. Sci.*, **44**, 1471-1479, 1989

[8] Mackley, M.R., Tweddle, G.M. and Wyatt, I.D., *Chem. Engng. Sci.*, **45**, 1237-1242, 1990

[9] Roberts, E.P.L., "Unsteady Flow and Mixing in Baffled Channels", PhD Thesis, Univ. of Cambridge, England, 1992

[10] Sobey, I.J., *J. Fluid. Mech.*, **96**, 1-26, 1980

[11] Brunold, C.R., Hunns, J.C.B., Mackley, M.R. and Thompson, J.W., *Chem. Engng. Sci.*, **44**, 1227-1244, 1989

[12] Rom-Kedar, V., Leonard, A. and Wiggins, S., *J. Fluid. Mech.*, **214**, 347-394 1990

Numerical Simulation of Turbulent Reacting Flows Using PDF Methods

R. O. Fox and M. R. Grier

Kansas State University, Manhattan, KS 66506

Turbulent reactive flow simulations are a challenging area of computational fluid dynamics (CFD) due to the wide range of length and time scales needed to correctly predict the reactive scalar fields. The well known difficulty of closing the reaction rate term limits the usefulness of standard turbulence models for reactive flow modeling. Probability density function (PDF) methods offer a tractable alternative approach that treats the single-point reaction rate terms exactly. The joint PDF balance equation is solved by a Monte-Carlo simulation. A hybrid PDF-CFD code, combining a CFD turbulence model to compute the mean velocity field and the turbulence fields with a PDF code to compute the velocity and reacting scalar fields, is briefly described.

Many reactive flows of interest in the chemical industry involve turbulent motion, complex chemical kinetics and geometrically nontrivial reactors. Commercially available turbulence codes, while often sufficient for computation of the velocity and turbulence fields, typically do a poor job of resolving reactive scalar fields when micromixing is a controlling factor (e.g., fast, nonpremixed second-order reactions). This fact is due to their inability to resolve (even by resorting to a prohibitively expensive ultrafine grid) the microscales of the scalar fields, especially at high Schmidt numbers. The goal of our research is thus to develop new computational fluid dynamics (CFD) codes for chemical reactor modeling that can accurately predict the composition fields in the presence of fast, finite-rate reactions. A computationally tractable approach for extending traditional CFD codes to reactive flow modeling can resort to so-called probability density function (PDF) methods (1,2).

With PDF methods both macromixing and micromixing are treated explicitly: macromixing is described by the mean velocity field, and micromixing is modeled by a subgrid-scale turbulent molecular mixing closure (1). This paper briefly describes work underway in our laboratory to develop and test new PDF codes. As with most PDF studies reported in the literature, our codes employ a standard CFD turbulence model to compute the mean velocity field (U), the turbulent kinetic energy field (k), and the turbulent energy dissipation field (ε); and a Monte-Carlo algorithm to compute the composition fields. Two types of PDF solvers are currently being investigated by our group: (a) a joint composition PDF code that models convective transport due to velocity fluctuations by gradient diffusion, and (b) a joint velocity-composition PDF code that includes velocity fluctuations and hence treats their contribution to convective transport exactly (1-3).

In the PDF approach, the crucial

unclosed term for reactive flows is that describing subgrid-scale turbulent molecular mixing. The most widely used closure for this term, mainly due to its computational efficiency, is the so-called IEM (or LMSE) model (1,4). Nevertheless, it is well known that the IEM model's prediction for the form of an inert scalar's PDF in isotropic turbulence is incorrect (4). Work by our group to solve this problem is underway and involves a stochastic description of diffusive turbulent mixing using a Fokker-Planck equation (5). In the calculations reported below, the IEM model has been employed; nevertheless, our PDF codes can easily be modified to incorporate new molecular mixing closures as they become available.

PDF METHODS

A one-point PDF description of a turbulent reactive flow can begin with the joint PDF of the velocity and compositions. This PDF is also a function of the position in the reactor for inhomogeneous flows, and of time for non-steady flows. Let u be a random variable denoting the instantaneous velocity and $\underset{\sim}{\phi}$ be a random vector denoting the compositions. The joint PDF of velocity and composition is then denoted by $f_{u,\underset{\sim}{\phi}}(v,\underset{\sim}{\psi};x,t)$ where

$$\int f_{u,\underset{\sim}{\phi}}(v,\underset{\sim}{\psi};x,t)\,dv\,d\underset{\sim}{\psi} = 1 \qquad (1)$$

i.e, the PDF integrates to unity. In these expressions, v is the sample-space variable for u, and $\underset{\sim}{\psi}$ for $\underset{\sim}{\phi}$.

Using the conservation equations for mass, momentum, enthalpy, etc., one can derive a single balance equation for the joint PDF (1). The general derivation is beyond the scope of this work; however, it should be noted that in order to arrive at a result useful for numerical simulations of real flows, closure approximations must be introduced. In general, a one-point PDF description is employed. Such a formulation contains no information about two-point correlations in the flow (e.g., the joint PDF at two different spatial locations in the flow at a given instant in time.) Physical processes involving gradients such as molecular diffusion and momentum transport thus lead to unclosed terms in a PDF description. However, terms involving single-point statistics, most notably the chemical reaction term, are treated exactly. Such single-point terms are precisely the ones that lead to closure problems in standard CFD codes when applied to reactive flows.

Joint Composition PDF

Many of the PDF codes employed to date utilize the joint composition PDF (6,7). This can be found from the joint velocity-composition PDF by integrating out the velocity variables:

$$f_{\underset{\sim}{\phi}}(\underset{\sim}{\psi};x,t) = \int f_{u,\underset{\sim}{\phi}}(v,\underset{\sim}{\psi};x,t)\,dv \qquad (2)$$

The balance equation for $f_{\underset{\sim}{\phi}}(\underset{\sim}{\psi};x,t)$ can be written as (1)

$$\frac{\partial}{\partial t}[\rho(\underset{\sim}{\psi})f_{\underset{\sim}{\phi}}] + \frac{\partial}{\partial x_i}[\rho(\underset{\sim}{\psi})U_i f_{\underset{\sim}{\phi}}] + \frac{\partial}{\partial \psi_\alpha}\left[\rho(\underset{\sim}{\psi})S_\alpha(\underset{\sim}{\psi})f_{\underset{\sim}{\phi}}\right] = \frac{\partial}{\partial \psi_\alpha}\left[\langle\frac{\partial}{\partial x_i}J_{i\alpha}(\underset{\sim}{\psi})|\underset{\sim}{\psi}\rangle f_{\underset{\sim}{\phi}}\right] - \frac{\partial}{\partial x_i}[\rho(\underset{\sim}{\psi})\langle u'_i|\underset{\sim}{\psi}\rangle f_{\underset{\sim}{\phi}}] \qquad (3)$$

wherein $\rho(\underset{\sim}{\psi})$ is the density, U_i the ith component of the mean velocity, $S_\alpha(\underset{\sim}{\psi})$ the reaction rate function for scalar α, $J_{i\alpha}(\underset{\sim}{\psi})$ the ith component of the diffusive flux for scalar α, and u'_i the fluctuating part of the ith component of the velocity. All terms on the left-hand side of the equals sign are closed and thus treated exactly. The two terms on the right-hand side of the equals sign involve unknown conditional expected values and thus must be modeled as described next.

The diffusive flux or turbulent molecular mixing term will be modeled in this work by the IEM model (1):

$$\langle \frac{\partial}{\partial x_i} J_{i\alpha}(\underset{\sim}{\psi}) | \underset{\sim}{\psi} \rangle = C_\phi \rho(\underset{\sim}{\psi}) \frac{\varepsilon}{k} (\psi_\alpha - \langle \phi_\alpha \rangle) \tag{4}$$

wherein $\langle \underset{\sim}{\phi} \rangle$ is the mean composition of scalar α, εk^{-1} the turbulent mixing rate, and $C_\phi = 1.0$ a "universal" mixing constant. Note that both the mean composition and the turbulent mixing rate may be functions of spatial position.

The convective flux due to turbulent velocity fluctuations is usually modeled by invoking a gradient-diffusion assumption (1,6,7). We shall proceed likewise. The turbulent diffusion closure thus has the form

$$\rho(\underset{\sim}{\psi}) \langle u'_i | \underset{\sim}{\psi} \rangle f_{\underset{\sim}{\phi}} = -\frac{\mu_T \partial f_{\underset{\sim}{\phi}}}{\sigma_T \partial x_i} \tag{5}$$

where the turbulent Schmidt number, σ_T, equals 0.7 and the turbulent viscosity is defined by

$$\mu_T = C_\mu \rho(\underset{\sim}{\psi}) \frac{k^2}{\varepsilon} \tag{6}$$

with $C_\mu = 0.09$.

Note that for constant density flows, $\rho(\underset{\sim}{\psi})$ can be eliminated from Equation 3. The PDF balance equation is then coupled to the CFD turbulence model through the mean velocity, U, and the turbulence fields, k and ε, appearing in the coefficients in Equations 4 and 6. For inhomogeneous turbulent flows, both k and ε will be functions of the spatial location in the flow. Thus both the rate of micromixing and the rate of turbulent diffusive mixing will depend on local turbulence quantities instead of global averages.

Joint Velocity-Composition PDF

A more complete description of a turbulent reactive flow can be obtained by solving the joint velocity-composition PDF balance equation (1,2,8):

$$\begin{aligned} &\frac{\partial}{\partial t}[\rho(\underset{\sim}{\psi}) f_{u,\underset{\sim}{\phi}}] + \frac{\partial}{\partial x_i}[\rho(\underset{\sim}{\psi}) v_i f_{u,\underset{\sim}{\phi}}] \\ &+ \frac{\partial}{\partial v_i}[\rho(\underset{\sim}{\psi}) g_i f_{u,\underset{\sim}{\phi}}] - \frac{\partial}{\partial v_i}\left[\left(\frac{\partial}{\partial x_i}\langle p \rangle\right) f_{u,\underset{\sim}{\phi}}\right] \\ &+ \frac{\partial}{\partial \psi_\alpha}\left[\rho(\underset{\sim}{\psi}) S_\alpha(\underset{\sim}{\psi}) f_{u,\underset{\sim}{\phi}}\right] = \\ &\frac{\partial}{\partial \psi_\alpha}\left[\langle \frac{\partial}{\partial x_i} J_{i\alpha}(\underset{\sim}{\psi}) | \underset{\sim}{\psi} \rangle f_{u,\underset{\sim}{\phi}}\right] \\ &+ \frac{\partial}{\partial v_i}\left[\langle -\frac{\partial \tau_{ji}}{\partial x_j} + \frac{\partial p'}{\partial x_i} | v, \underset{\sim}{\psi} \rangle f_{u,\underset{\sim}{\phi}}\right] \end{aligned} \tag{7}$$

wherein the closed terms in order of appearance are (i) accumulation, (ii) convective flux in real space due to velocity, (ii) flux in velocity space due to body forces, (iv) flux in velocity space due to mean pressure gradient, and (v) flux in composition space due to reactions; and the unclosed terms are (vi) flux in composition space due to turbulent molecular mixing, and (vii) flux in velocity space due to viscous stresses and fluctuating pressure gradient. Note that spatial transport due to velocity fluctuations is treated exactly (term ii) and thus no gradient-diffusion approximation is needed.

The new unclosed term (vii) will be modeled here using the generalized Langevin model (9,10):

$$\begin{aligned} &\langle -\frac{\partial \tau_{ji}}{\partial x_j} + \frac{\partial p'}{\partial x_i} | v, \underset{\sim}{\psi} \rangle f_{u,\underset{\sim}{\phi}} = \\ &- G_{ij}(v_j - U_j) + C_0 \varepsilon \frac{\partial f_{u,\underset{\sim}{\phi}}}{\partial v_i} \end{aligned} \tag{8}$$

wherein G_{ij} is a second-order tensor of local mean quantities, and C_0 is a positive constant. For homogeneous isotropic turbulence, G_{ij} must be isotropic with a coefficient of proportionality determined by requiring that the modeled equation yield the correct energy decay rate (1):

$$G_{ij} = \left(\frac{1}{2} + \frac{3}{4}C_0\right)\delta_{ij} \qquad (9)$$

wherein C_0 equals 2.1. A more general model for the case of homogenous anisotropic turbulence with uniform mean velocity gradients is also available (9).

In applications, a CFD code is needed to predict the local turbulent energy dissipation, ε, appearing in Equation 8. The CFD code also supplies values for the mean velocity, U, and the turbulent kinetic energy, k, which should correspond to the first and second moments of the velocity computed by the PDF code. In order to ensure that this is the case, Equation 8 is modified to include a translation (mean correction) and scaling factor (variance correction) at each time step.

SOLUTION ALGORITHM

The simplest inhomogeneous flow is one involving passive scalars, e.g., scalars that do not affect the fluid dynamics. For this case, the flow field can be computed separately from the scalar fields so that the CFD code will be uncoupled from the PDF code. Only this case will be considered here. Nonconstant density cases and active scalars have been considered elsewhere (6,7,8,11). As compared to the passive scalar case, computational requirements for the active scalar case are significantly greater. Nevertheless, the basis computational methodology is identical for both cases.

The probability balance appearing in Equation 7 involves at least six independent variables for a two-dimensional flow with one scalar component. For complicated kinetic expressions, several scalars will generally be needed, drastically increasing the dimensionality of the problem. It should thus be clear that standard discretization methods for partial differential equations will be intractable. Instead, a Monte-Carlo simulation algorithm can be employed by replacing the joint PDF with an ensemble of notional particles. As a rule of thumb, ensemble sizes should be on the order of 300 notional particles per finite-volume cell used in the CFD code in order to achieve good statistical estimates. Note that an ensemble with a single notional particle at each cell would correspond to a finite-volume calculation of the scalar fields (i.e., the scalar field at each grid point is represented by a single number). The PDF code thus gives a more detailed representation of the scalar field that includes deviations about the mean value.

In the Monte-Carlo algorithm, each notional particle in the ensemble carries with it a set of variables which includes:

- x-coordinate position: x_1
- y-coordinate position: x_2
- z-coordinate position: x_3
- x velocity: u_1
- y velocity: u_2
- z velocity: u_3
- scalar 1: ϕ_1
- scalar 2: ϕ_2
- ...

These variables must be updated at each time step in the simulation according to the terms appearing in the PDF balance equation (e.g., Equation 3 or 7). Due to the nature of the balance equation (and the closure models), it can be replaced by an equivalent set of stochastic differential equations. The latter can be solved using standard techniques (1).

For example, for a constant density flow Equation 7 leads to the following set of stochastic differential equations involving the variables of the kth notional particle:

$$dx_i^{(k)} = u_i^{(k)}dt \qquad (10)$$

(Change in spatial location.)

$$du_i^{(k)} = \left(g_i - \frac{1}{\rho}\frac{\partial}{\partial x_i}\langle p\rangle\right)_{x_i = x_i^{(k)}} dt + \frac{1}{\rho}G_{ij}\left(U_j - u_j^{(k)}\right)_{x_i = x_i^{(k)}} dt + \sqrt{2C_0\frac{1}{\rho}\varepsilon}\, dW_{ui}(t) \qquad (11)$$

(Change in velocity.)

$$d\phi_\alpha^{(k)} = S_\alpha(\underset{\sim}{\phi}^{(k)})\,dt + C_\phi\frac{\varepsilon}{k}\left(\langle\phi_\alpha\rangle - \phi_\alpha^{(k)}\right)dt \qquad (12)$$

(Change in scalar composition.)

The symbol $dW_{ui}(t)$ denotes a Wiener process with zero mean and unit variance (1). In the Monte-Carlo simulation, it is approximated by

$$dW(t) \approx \xi\sqrt{dt} \qquad (13)$$

where ξ is a standard normal random number.

In Equation 11, the subscript $x_i = x_i^{(k)}$ indicates that the expression in parenthesis is to be evaluated at the spatial location of the *k*th notional particle. Although not shown explicitly, this is also the case for the turbulence coefficients and the expected values in Equations 11 and 12. In the simulation code, this information is provided either by a smoothing spline fitted using the notional particle values as input data, or by a local averaging process using the notional particle values in the vicinity of the *k*th notional particle.

The basic algorithm for simulating the joint PDF balance equation for a constant density, passive scalar, turbulent flow thus proceeds as follows

a) In a presimulation, compute mean velocity and turbulence fields in the reactor using a CFD code.

b) Initialize variables for all notional particles.

c) Advance variables one time step using Equations 10-12.

d) Check boundary conditions, adding new notional particles at the inlet to replace those that move out of reactor.

e) Repeat c and d until desired stopping time is attained.

The boundary conditions required in d are for the physical boundaries of the reactor. At solid boundaries, reflecting boundary conditions are typically employed for x_i and u_i. At outflow boundaries, notional particles are removed from the system. The scalar values of outflowing notional particles can be used to estimate the outflow concentrations. Finally, notional particles added at inflow boundaries are assigned scalar compositions and velocities consistent with the inlet stream.

NUMERICAL EXAMPLE

A hybrid CFD-PDF code has been applied to study an axially symmetric expanding, coflowing turbulent jet. In a presimulation, a CFD code (FLUENT) was employed to compute the turbulence fields needed for the PDF code. The inlet jet was isokinetic; thus a uniform inlet velocity was employed. The physical dimensions of the jet are shown in Figure 1. The jet flows from left to right in the figure, entering on the left through the region from 0.1 to -0.1 and exiting on the right through the region from 0.2 to -0.2. The inlet stream carries a nonpremixed, inert passive scalar (e.g., a tracer) with value 1.0 up to a radial distance of 0.05 from the centerline and 0.0 elsewhere. The physical properties of water were used. Various reaction schemes have been investigated; however, due to lack of space only results from the inert scalar case will be shown here.

Both a composition PDF code and a joint velocity-composition PDF code have been tested. For the inert scalar case, results from both codes are nearly identical. The mean scalar field is shown in Figure 1 and the variance of the scalar field in Figure 2. Nonzero variance occurs in regions where micromixing is a controlling factor. As can be seen from Figure 2, these regions occur near the inlet where the feedstream is highly segregated. Mixing in this flow is strongly dependent on turbulent velocity fluctuations which transport fluid elements radially. The mean scalar field, which for an inert scalar is independent of the micromixing rate, is determined by turbulent convective transport. The rate of this process varies continuously with spatial location in the flow according to the computed values of the turbulence fields. In general, turbulent diffusive mixing is strongest near the core of the jet and near the inlet.

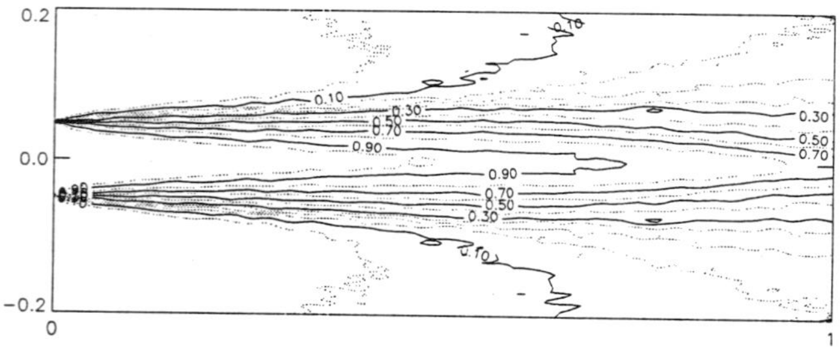

Figure 1. Computed scalar mean field.

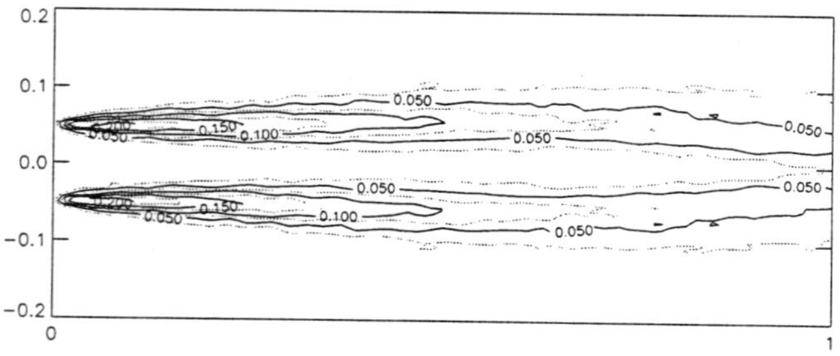

Figure 2. Computed scalar variance field.

ACKNOWLEDGEMENTS

This work is supported by the National Science Foundation under grant CTS-9158124 (PYI Award). The authors wish to thank Prof. S. B. Pope for use of his joint velocity-composition PDF code (pdf2ds).

LITERATURE CITED

1. Pope, S. B., Prog. Energy Combust. Sci., **11**, 119 (1985).

2. Pope, S. B., "Computations of turbulent combustion: Progress and challenges," *23rd International Symposium on Combustion*, (The Combustion Institute, Pittsburgh, 1990), pp. 591-612.

3. Fox, R. O., Chem. Engng. Sci., **47**, 2853 (1992).

4. Fox, R. O., "The Fokker-Planck molecular mixing closure for PDF modeling of turbulent flows," AIChE Symp. Ser. **88**, 286, 83 (1992).

5. Fox, R. O., Phys. Fluids A **4**, 1230 (1992).

6. Roekaerts, D., Appl. Scientif. Res. **48**, 271 (1991).

7. Chen, J.-Y., W. Kollman and R. W. Dibble, Combust. Sci. and Tech., **64**, 315 (1989).

8. Pope, S. B., Combust. Sci. and Tech., **25**, 159 (1981).

9. Haworth, D. C. and S. B. Pope, Phys. Fluids **29**, 387 (1986).

10. Pope, S. B., Phys. Fluids **24**, 588 (1981).

11. Haworth, D. C. and S. H. El Tahry, AIAA J. **29**, 208 (1991).

Bifurcations and Mixing Windows in Piecewise Steady Mixing Systems

F. H. Ling

Polymer Processing Institute and Department of Chemistry and Chemical Engineering
Stevens Institute of Technology, Hoboken, NJ 07030

The islands around stable periodic orbits of low period are the main obstacles to global chaos and hence to uniform mixing of fluids. The periodic orbits of piecewise steady mixing systems can be constructed geometrically as a composition of segments of streamlines belonging to different steady states, and some critical orbits can be analyzed semi-analytically. Starting from these critical orbits, one can follow a family of periodic orbits and determine their stability. Mixing windows which lead to uniform mixing in the parameter space are then found by excluding stable regions of these orbits. Mixing in counter-rotating discontinuous cavity flows are studied in detail to illustrate the strategy.

1 Introduction

Mixing of fluids in laminar flows is involved in a variety of natural and engineering processes, such as in food and polymer processing and in the mantle of the earth [1]. Recently, there has been extensive interest in exploring techniques from chaos theory, in particular Poincaré sections and fingerprints of local chaos, to study mixing in incompressible and creeping flows [1-7]. While in most applications one tries to avoid chaos, chaos is desired in mixing. More precisely, nearly global chaos is often a necessary condition for producing uniform mixtures. Therefore, the presence of fingerprints of local chaos, such as homoclinic and heteroclinic orbits, period-doubling, intermittency, etc. is—though interesting from a theoretical point of view—not informative for practical applications. Numerical simulations, such as generating Poincaré sections and tracking deformation of a blob to be mixed by discretizing them into many particles, are useful but are time-consuming, and lack insight into the underlying mechanism, and hence, cannot give useful hints on how to create global chaos and hence, uniform mixing. Noting that the islands around stable low-periodic orbits are the main obstacles to global chaos, we concentrate on the periodic orbits instead of chaos. If all periodic orbits become unstable, the system is globally chaotic. More practically, if all periodic orbits with measurable islands become unstable, the system is nearly globally chaotic and satisfies practical demands [8].

Based on this, the concept of *mixing windows* has been proposed [9,10]. The mixing windows are defined as regions in the parameter space, where the mixing system is globally chaotic. In other words, no periodic orbit is stable if parameters of a mixing system fall into the mixing windows. It is noticed that higher periodic orbits might exist for all parameters, but most of them do not possess measurable islands and can be omitted. The mixing windows in (continuous) sinusoidal cavity flows are found by tracking low-periodic orbits using a shooting type numerical method [9] and in the co-rotating discontinuous cavity flow semi-analytically with help of a geometrical construction of periodic orbits [10]. The second approach is more efficient and can assure finding a class of periodic orbits in piecewise steady mixing systems, whose initial point lies on the axis of symmetry. Many important periodic orbits do not belong to this class except for some of their critical configurations. These critical orbits can be analyzed semi-analytically, and starting from them, one can follow the periodic orbits in their families and analyze the stability and hence, find mixing windows.

Piecewise Steady Planar Flow and Discontinuous Cavity Flow

Mixing is characterized by the passive motion of the particles in such flows, described by a Lagrangian system,

$$\begin{cases} \dot{x} = u_i(x,y) \\ \dot{y} = v_i(x,y) \end{cases}, \quad i = \begin{cases} 1 & \text{for } 0 \leq t < \alpha T \\ 2 & \text{for } \alpha T \leq t < T \end{cases}, \tag{1}$$

where T is the period of the motion of the boundaries, $0 \leq \alpha \leq \frac{1}{2}$. The streamlines are symmetric about the y-axis, i.e., $u_i(-x,y) = u_i(x,y)$, $v_i(-x,y) = -v_i(x,y)$, $i = 1,2$, and so do the periodic orbits of (1).

We turn now to a typical piecewise steady mixing system—the discontinuous cavity flow mixing system. The cavity studied is a rectangle with aspect ratio h(height)/w(width)=0.6; its top and bottom walls move alternately and periodically with the same constant speed, $v_{wall} = 0.2612\ w/second$. The flow is called co-rotating if the motion of both walls leads to fluid rotation in the same direction (e.g., clockwise) and counter-rotating if they induce rotation in the opposite directions. The co-rotating flow is induced when the top and bottom walls move in the opposite direction, while the counter-rotating flow is produced by both walls moving in the same direction. We will mainly discuss the counter-rotating flow in this paper, but some new results on the co-rotating flow will also be presented. The velocity field is calculated using a finite difference method and spline interpolation. Newtonian fluids are considered in this paper, but an extension to simple non-Newtonian fluids, such as generalized Newtonian fluids, is straightforward.

We first consider a special case, where the translation of both walls is identical during one period, i.e., $\alpha = \frac{1}{2}$ in (1). The streamlines of the bottom wall flow are symmetric to those of the top wall flow about the x-axis. It was shown that except for the symmetry about the y-axis, periodic orbits in this system are either self-symmetric or mutually (in pairs) symmetric about the x-axis [9].

Geometrical Construction of Periodic Orbits

The periodic orbits of (1) consist of segments of the streamlines of the top and bottom wall flows. There is a simple critical case: the periodic orbit consists of two identical complete loops connected at the center of the cavity as shown in Figure 1 (a). There a dot denotes the beginning as well as the end of the periodic orbit, and a cross its middle, where the motion of the walls switches. The corresponding period is denoted with T_{11}. If the period of the walls, T, is shifting from T_{11}, two different kinds of P-1 orbits are possible as shown in Figure 1 (b) and (c), denoted with P-1N_{11} (normal loop) and P-1D_1 (double loop), respectively [14]. Two arrows in Figure 1 (b) indicate that the corresponding segment is passed twice during one period. From above, the critical periodic orbit is not structurally stable, and T_{11} is a bifurcation value [11]. Similar critical orbits emerge at $T_{nn} = n \cdot T_{11}$, which have the same form as that at T_{11}, but the two loops are passed each n times during one period. The periodic orbits P-1N_{nn} and P-1D_n are formed similarly when T shifts from T_{11}. P-1D_n is self-symmetric about the x-axis, and P-1N_{nn} appear in mutually symmetric about the x-axis pairs.

Other kinds of critical orbits exist, where a streamline loop in the upper wall flow is passed m times, and a different loop in the bottom wall flow is passed n times ($m \neq n$), during one period. Figure 1 (d) shows the case for $m = 2$ and $n = 1$; the corresponding bifurcation value is $T = T_{21}$. Further P-1 orbits will form when T shifts from T_{21} as shown in Figure 1 (e), denoted with P-1N_{21} which has a symmetric companion P-1N_{12}. Since both walls move the same distance during one period, we have $T_{12} = T_{21}$. One can form the periodic orbits P-1N_{mn} in a similar manner. In addition, there are P-1N_{10} and P-1N_{01} originating from the $T \to 0$ limit as shown in Figure 1 (f).

Critical Orbits and Bifurcation

P-1N_{10} and P-1N_{01} degenerate to the stagnation points of the flow generated by two continuously moving walls —the $T \to 0$ limit of the discontinuous flow [9]. Bifurcation values T_{mn} can be determined based on the cycling time of the points on the y-axis. Denoting $T_{cyc}(y)$ the cycling time of the point $(0,y)$ along a streamline loop of the top wall flow and noticing that the cycling time of the same point along a streamline loop of the bottom wall flow is $T_{cyc}(-y)$, one can determine y_{bif} from

$$m \cdot T_{cyc}(y) - n \cdot T_{cyc}(-y) = 0, \tag{2}$$

and the bifurcation value is $T_{mn} = 2m \cdot T_{cyc}(y_{bif})$ as shown in Figure 2 graphically. There the lowest solid

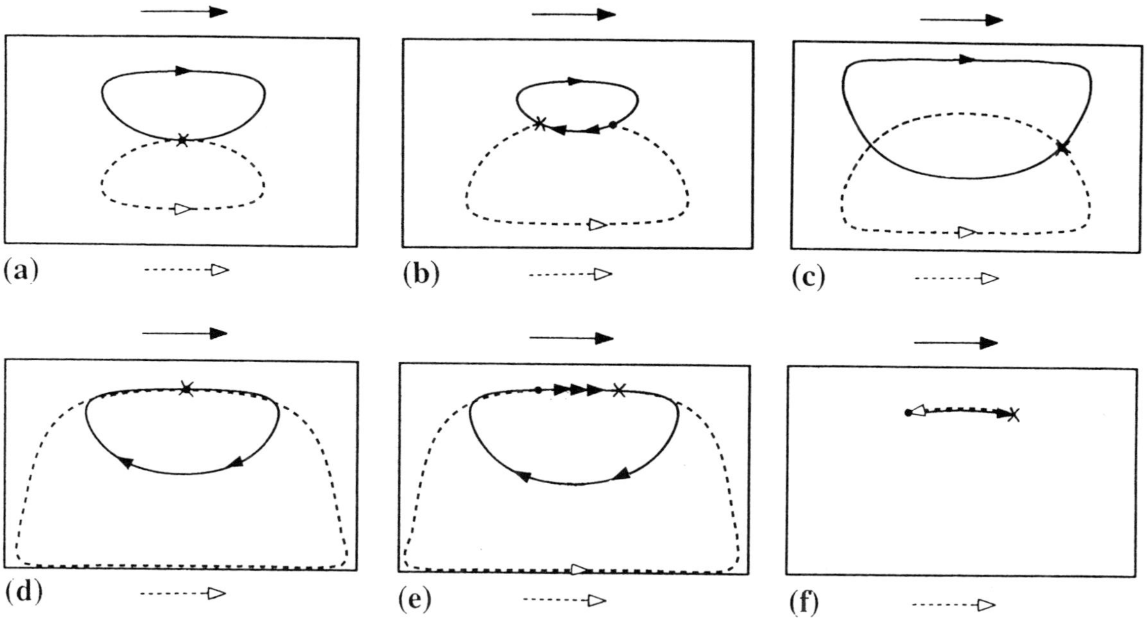

Figure 1: Basic kinds of P-1 orbits of the counter-rotating discontinuous cavity flow mixing systems, (a) critical orbit of P-1N_{11} and P-1D_1, (b) normal loop, P-1N_{11}, (c) double loop, P-1D_1, (d) critical orbit of P-1N_{21}, (e) normal loop P-1N_{21}, (f) normal loop P-1N_{10}.

and dashed lines represent $T_{cyc}(y)$ and $T_{cyc}(-y)$, respectively. The other lines are $kT_{cyc}(y)$ and $kT_{cyc}(-y)$, $k = 2, 3, 4, 5$. A few smallest T_{mn} and corresponding P-1 points of the induced Poincaré map [12], y_{mn}, are given in Tab. 1. Stagnation points in the co-rotating flow (with index 00) and counter-rotating flow (with index 01 and 10) are also listed in Tab. 1.

Table 1 Some bifurcation points of discontinuous cavity flow mixing maps

m	n	T_{mn}	y_{mn}	m	n	T_{mn}	y_{mn}
0	0	0	0	1	2	71.77	0.2156
1	0	0	0.1734	3	3	106.85	0.1985
0	1	0	-0.1734	3	2	107.21	-0.1796
1	1	35.62	0.1985	2	3	107.21	0.0276
2	2	71.23	0	3	1	116.06	0.2956
2	1	71.77	0.2901	1	3	116.06	0.0910

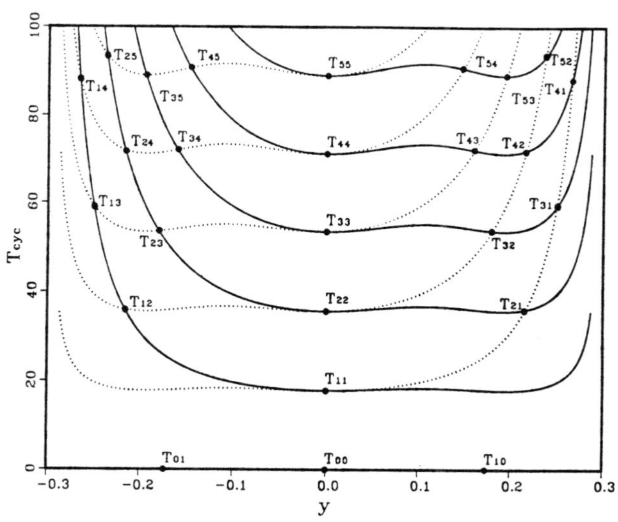

Figure 2: Évaluation of bifurcation values of the discontinuous cavity flow mixing system with $\alpha = \frac{1}{2}$.

It is noticed that the critical orbits are the same regardless the directions of the motion of the walls. Therefore, the bifurcation values obtained above are for both co- and counter-rotating flows. As seen above, there are two sequences of the bifurcation values; one is periodic, T_{nn}, and another is non-periodic, $T_{mn}(m \neq n)$. The first one has been discovered for the co-rotating flow using a different method [10].

Stability and Mixing Window

The P-1 orbits in the counter-rotating flow were tracked starting from the critical orbits [13]; see Figure 3 for a bifurcation diagram of the corresponding Poincaré map [14]. Born at $T = 0$, P-1N_{10} undergoes a period-doubling bifurcation at $T \approx 13.2$, and the resulting P-2 orbit bifurcates to a P-4 orbit at $T \approx 25.8$, etc., leading to chaos shortly after that. P-1N_{01} bifurcates in the same way as it should be, but the resulting P-2 points are out of the y-axis. The pair of P-1N_{11} is born stable at $T_{11} \approx 35.62$ together with an unstable pair which is symmetric to P-1N_{11} about the y-axis. P-1N_{11} undergo an infinity of period-doubling bifurcations starting at $T \approx 46.5$. The bifurcation schema of other P-1N_{nn} is similar, e.g., the stable region of P-1N_{22} is $T \in (73.23, 80.1)$ approximately. P-1D_n are born at T_{nn} and are always marginally stable. They have islands only in a small T-interval, which is mostly overlapped with the stable region of P-1N_{nn}. P-1N_{21} undergoes a pitchfork bifurcation at $T_{21}(\approx 71.7)$, becomes unstable for a increasing T and generates two stable P-1 orbits, which are mutually symmetric about the y-axis. It becomes stable again after a (reverse) pitchfork bifurcation at $T \approx 77.75$ and goes to chaos through an infinite sequence of period-doubling bifurcation starting at $T \approx 80.9$. Tracking it back, we found that it is born at $T \approx 64.7$ through a tangent bifurcation. P-1N_{32} also undergoes a pitchfork bifurcation at $T_{32}(\approx 107.2)$, but becomes stable for an increasing T and bifurcates to a P-2 at $T \approx 115.0$. Tracking it back, we found that it is born unstable at $T \approx 92.6$ through a tangent bifurcation. The stable P-1 born together loses its stability almost immediately at $T \approx 92.68$.

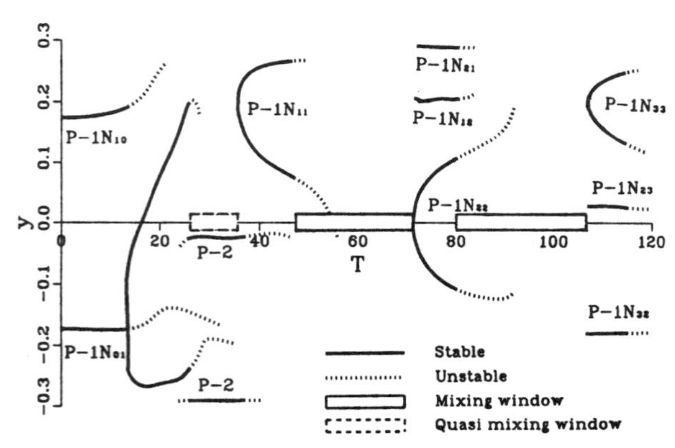

Figure 3: Bifurcation diagram of the counter-rotating discontinuous cavity flow mixing system with $\alpha = \frac{1}{2}$.

Besides, there is a stable P-2 pair in $T \in (25.69, 36.6)$ approximately, but the islands around this P-2 orbit are small as seen in the Poincaré sections calculated.

If one cuts off the T-regions where at least one of the above-mentioned P-1 or P-2 orbits is stable, the remains is the mixing windows. From Figure 3, the first mixing window is at (46.5, 64.7) approximately, the second one at (80.7, 107.0) approximately, etc. The interval (25.8, 35.7) may be called a quasi mixing window, since there are only very small islands. To verify, we show the Poincaré sections in the quasi mixing window, $T = 30$, Figure 4 (b), in the mixing window, $T = 56$ and 100, Figure 4 (c) and (e), and outside the mixing window T =12 and 78, Figure 4 (a) and (d).

Stability of the P-1 orbits originating from the periodic sequence of bifurcation values of the co-rotating flow has been determined earlier [10]. An investigation of the periodic orbits originating from the non-periodic sequence of bifurcation values in the co-rotating flow reveals that their stability regions are mostly overlapped with those of the periodic sequence. The first three mixing windows in the co-rotating flow are at (19.13, 22.00), (49.43, 61.30), and (83.60, 98.16) approximately. Including the quasi mixing window, the number of mixing windows is the same for both flows, but the windows in the counter-rotating flow are wider. This kind of behavior is in contrast to that of sinusoidal (continuous) cavity flows (where two walls move sinusoidally with a different initial phase), where the counter-rotating flow is apparently superior in terms of the mixing performance. This is caused by the existence of an adiabatic limit of the co-rotating sinusoidal (and other continuous) cavity flow mixing systems at $T \to \infty$ [9].

Remarks

Bifurcation values of general discontinuous cavity flows ($\alpha \neq \frac{1}{2}$ in (1)) can be similarly calculated as $T_{mn} = m \cdot T_{cyc}(y_{bif})/\alpha$, where y_{bif} is determined from

$$\frac{m}{\alpha} \cdot T_{cyc}(y) - \frac{n}{1-\alpha} \cdot T_{cyc}(-y) = 0. \qquad (3)$$

It is clear that only $\alpha \leq \frac{1}{2}$ needs to be considered. The bifurcation sets in (T, α) space are calculated and plotted in Figure 5. A decline of α from $\frac{1}{2}$ breaks up the symmetry between the top and bottom wall flows; the periodic sequence of bifurcation values becomes non-periodic, and two originally overlapped non-periodic sequences split into two different sequences. Hence, total number of the bifurcation values increases, so that

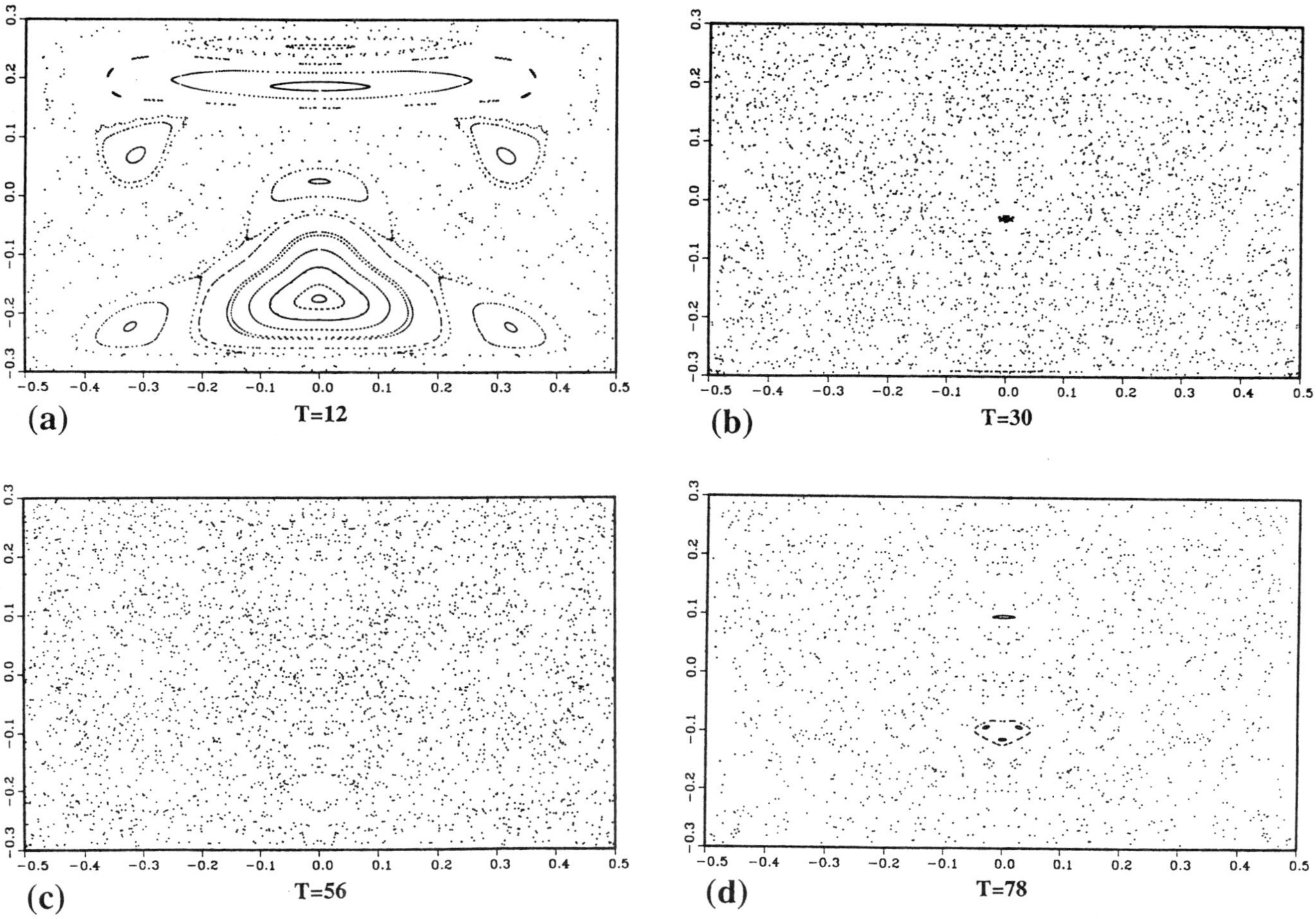

Figure 4: Poincaré sections of the counter-rotating discontinuous cavity flow mixing system with $\alpha = \frac{1}{2}$, (a)$T = 12$, (b)$T = 30$, (c)$T = 56$, (d)$T = 78$.

a decline of α from $\frac{1}{2}$ may worsen mixing performance. On the other hand, the smallest bifurcation value increases with a decreasing α, and this may widen the first mixing window. Nevertheless, the Lagrangian system (1) is integrable for $\alpha = 0$, so we do not expect a noticeable improvement of the mixing performance when α declines from $\frac{1}{2}$. On the contrary, breakup of the symmetry has a significant effect on the co-rotating sinusoidal cavity flow, where the mixing performance of the symmetric case with a phase shift of $T/2$ is very bad, while a phase shift of $T/4$ deems to give the best results [9].

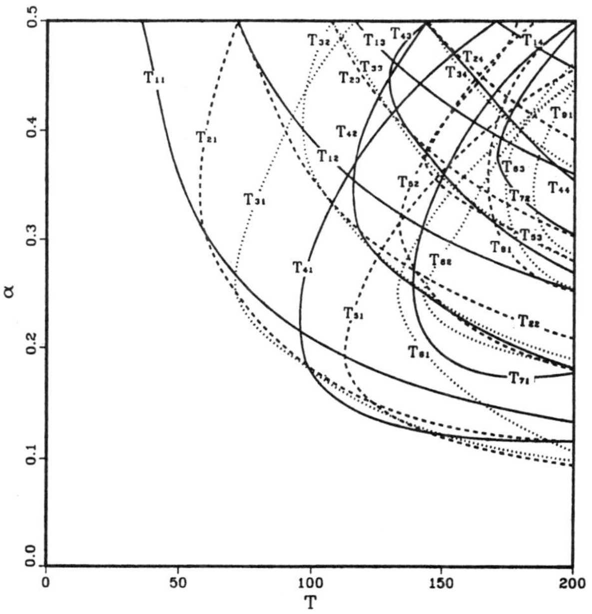

Figure 5: Bifurcation sets of the discontinuous cavity flow mixing systems.

Mixing in other piecewise steady planar flows, such as journal bearing flows and blinking vortex flows will be studied using the methodology given in this paper as well. It is also noticed that the three-dimensional steady flow periodic in one spatial direction can be treated similarly, if one regards the coordinate in this direction as the time t in this paper; see ref. [15] for an example.

Acknowledgements

This work is part of the industrially co-sponsored "Polymer Mixing Study" of the Polymer Processing Institute. The author thanks G. Schmidt for discussions and X. Zhang for his assistance in drawing graphs.

Reference

[1] J. M. Ottino, *The Kinematics of Mixing: Stretching, Chaos, and Transport* (Cambridge University Press, Cambridge, 1989); J. M. Ottino, *Annu. Rev. Fluid Mech.* **22**, 207 (1990); Proceedings of the IUTAM Symposium of Fluid Mechanics of Stirring and Mixing, *Phys. Fluids* **A 3** No. 5, Pt. 2 (1991).

[2] H. Aref, *J. Fluid Mech.* **143**, 1 (1984); S. W. Jones, *Phys. Fluids* **A 3**, 1081 (1991).

[3] D. V. Khakhar, R. Rising and J. M. Ottino, *J. Fluid Mech.* **172**, 419 (1986).

[4] W.-L. Chien, H. Rising and J. M. Ottino, *J. Fluid Mech.* **170**, 355 (1986); J. G. Franjione, C. W. Leong and J. M. Ottino, *Phys. Fluids* **A 1**, 1772 (1989); C. W. Leong and J. M. Ottino, *J. Phys. Fluids* **A 1**, 1441 (1989); C. W. Leong, A Detailed Analysis of Chaotic Mixing of Viscous Fluids in Time-Periodic Cavity Flows, Ph. D. Thesis, University of Massachusetts, Amherst (1989); M. Liu and R. L. Peskin, *Phys. Fluids* **A 3**, 1436 (1991).

[5] C. W. Leong and J. M. Ottino, *J. Fluid Mech.* **209**, 463 (1989).

[6] H. Aref and S. Balachandar, *Phys. Fluids* **29**, 3515 (1986); J. Chaiken, R. Chevray, M. Tabor and Q. M. Tan, *Proc. R. Soc. Lond.* **A 408**, 165 (1986); J. Chaiken, C. K. Chu, M. Tabor and Q. M. Tan, *Phys. Fluids* **30**, 687 (1987); C. W. Leong and J. M. Ottino, *J. Phys. Fluids* **A 2**, 1524 (1989); P. D. Swanson and J. M. Ottino, *J. Fluid Mech.* **213**, 227 (1990); H. Aref, *Phys. Fluids* **A 3**, 1009 (1991); F. J. Muzzio, P. D. Swanson and J. M. Ottino, *Phys. Fluids* **A 3**, 882 (1991).

[7] S. W. Jones and H. Aref, *Phys. Fluids* **31**, 469 (1988); S. W. Jones, O. M. Thomas and H. Aref, *J. Fluid Mech.* **209**, 335 (1989); T. J. Danielson and J. M. Ottino, *Phys. Fluids* **A 2**, 2024 (1990); S. M. Cox, P. G. Drazin, S. C. Ryrie and K. Slater, *J. Fluid Mech.* **214**, 517 (1990); F. Városi, M. A. Thomas, Jr. and E. Ott, *Phys. Fluids* **A 3**, 1017 (1991); D. Beigie, A. Leonard and S. Wiggins, *Phys. Fluids* **A 3**, 1039 (1991); J M. Ottino, *Phys. Fluids* **A 3**, 1417 (1991); D. V. Khakhar and G. Sridhar, *Phys. Fluids* **A 3**, 1437 (1991); D. Barkley and G. E. Karniadakis, *Phys. Fluids* **A 3**, 1063 (1991).

[8] Locating periodic orbits is a common practice in analysis using chaos theory; this was also carried out for some mixing systems [3,5]. On the other hand, there were rare attempts to find nearly globally chaotic systems using this knowledge, but see F. H. Ling, *Z. für Phys. B* **75**, 539 (1989) and F. H. Ling and Y. M. Cheung, in *Nonlinear Dynamics in Engineering Systems* (ed. W. Schiehlen, Springer Verlag, Berlin, 1990) for such a treatment of dissipative systems.

[9] F. H. Ling and G. Schmidt, Bifurcations and Mixing Windows in Sinusoidal Cavity Flow Mixing Systems, Preprint.

[10] F. H. Ling and G. Schmidt, *Phys. Lett. A* **165**, 221 (1992).

[11] J. Guckenheimer and P. Holmes, *Nonlinear Oscillations, Dynamical Systems, and Bifurcations of Vector Fields* (Springer-Verlag, New York 1983), p. 119.

[12] Suitable choice of the sampling time will assign the resulted Poincaré map a reversibility corresponding to one of the symmeties of the periodic orbit [9]. We choose the starting sampling time at $\frac{1}{2}\alpha T$, so that the y-axis is always the axis of symmetry of the Poincaré map.

[13] One can use a standard shooting type method, or better, match traveling time along the streamlines of the top and bottom wall flows. The latter one will be reported in detail elsewhere.

[14] The $T_{cyc} - y$ curve happens to have a minimum almost at $y = 0$ for this particular case, $h/w = 0.6$. Therefore, if T decreases from T_{nn}, P-1N$_{nn}$ exist only for a small T-interval. On the other hand, P-1D$_n$ exist only for $T > T_{nn}$ as can be figured out from its geometric configuration.

[15] F. H. Ling, Bifurcations and Mixing Windows in Spatially Periodic Continuous Mixers, Preprint.

Numerical Simulation of Turbulent Flow In Bubble Column Reactors

Vivek V. Ranade

Chemical Engineering Division, National Chemical Laboratory, Pune 411 008, India

A numerical simulation model for predicting the details of turbulent gas-liquid flow in bubble columns is presented. Influence of sparger characteristics on flow characteristics are investigated in detail for the first time. Major issues in the flow modelling of bubble columns are discussed.

Bubble columns are being widely used in the chemical and petrochemical industry. The development of accurate models for scaling up and designing of these reactors would require detailed knowledge of the flow and mixing characteristics of the reactor. Most of the published mixing models make use of axial dispersion coefficient or tanks in series concepts, disregarding the actual details of the flow. It is however necessary to develop mixing models based on the turbulent fluid mechanics of bubble column reactors to make them more realistic.

Fluid mechanics of bubble columns has been the topic of a large number of publications in the last two decades. However, no general flow model has been developed until now. This is because of the lack of understanding about complicated interactions between numerous hydrodynamic parameters. Internal liquid circulation due to the non-uniform radial holdup profile is a commonly encountered flow structure in bubble columns. Though the importance of internal liquid circulation is recognised, no theory based on first principles is available. Most of the published flow models are all restricted to one dimensional approximations which require the radial gas holdup profile and eddy viscosity as model inputs. In this paper we have attempted to address this problem using CFD approach for simulating the fluid mechanics of bubble column reactors.

PREVIOUS WORK

Recently, detailed experimental data about flow in bubble columns have been published (Devanathan et al. [1]; Yao et al. [2]; Menzel et al. [3]; Torvik [4]; Yu and Kim [5]; Gross and Kuhlman [6]), which clearly indicate the complex interactions of bubble slip velocity, radial holdup profile and internal circulation in bubble columns. For the interpretation and understanding of these data, more advanced models dealing with more than one spatial dimension are necessary. Recently, computational fluid dynamics (CFD) based approach is being adopted to understand the fluid mechanics of the bubble columns. Torvik and Svendsen [7], Ranade [8], Netti and Mohamed [9], Grienberger and Hofmann [10], Svendsen et al. [11] and Webb et al. [12] have published initial results of such CFD approach.

Neti and Mohamed [9] and Webb et al. [12] have used the Lagrangian approach to individual bubbles in a Eulerian liquid framework. However, extension of this approach for high throughput of gas will be difficult and will be more computation intensive. For the simulation of flow in bubble columns, use of Eulerian framework for both the dispersed and continuous phase would be more suitable. Torvik and Svendsen [7] have attempted to use such framework to simulate three phase slurry reactors. They have used standard k-ϵ model of turbulence. Recently, Svendsen et al. [11] have introduced an additional source term in the equation of turbulent kinetic energy to account for the presence of bubbles. They report reasonable agreement with the experimental data not only for the mean velocities but also for the turbulent kinetic energy. Grienberger and Hofmann [10] have used standard k-model. Both of these studies use lift coefficient as a tuning parameter to get the desired predictions (Svendsen et al. have used -3 and Grienberger and Hofmann have used -0.5). In addition to this, Grienberger and Hofmann also use two empirical relations (one for the radial profile of the drag coefficient and other for the effective turbulent viscosity of the gas phase). Both of these groups concentrate more on showing the possibility of fitting the experimental data with the model rather than developing an understanding about the influence of various factors on the flow in bubble columns. The flow in bubble columns is strongly influenced by the sparging patterns and the sparger characteristics. The existence of non-uniform radial profile of gas holdup in the column will eventually lead to the non-uniform gas distribution at the sparger. However, none of the previous attempts have considered this aspect. Both, Grienberger and Hofmann [10] and Svendsen et al. [11] have reported model predictions for the column with fixed height to diameter ratio (H/D=10 or so). No discussion is offered on the influence of height to diameter ratio on predicted flow characteristics. Both of these studies report comparison with experimental data only at one height. However, whether the shown agreement holds good for all the heights is not reported. The vector plot reported by Svendsen et al. for bubble column indicate the radially inward flow only for the bottom most level. All the levels above it show radially outward flow. This overall pattern does not agree with the patterns reported by Devanathan [1]. Considering these aspects it is worthwhile to investigate the detailed predictive characteristics of the flow model. Recently, Ranade et al. [8] has attempted such investigation by means of numerical experiments. However, a simplified flow model was used in that work. It is necessary to continue such investigation to understand the behaviour of flow models for bubble columns. In this paper, we have undertaken this task using better flow model.

MATHEMATICAL MODEL

The case of gas-liquid flow in cylindrical bubble column is considered. The cylindrical co-ordinate system has been used with the origin located at the bottom centre. A schematic diagram of the column with the co-ordinate system is shown in Figure 1. In order to highlight the key features of internal circulation in bubble columns, we have eliminated non-essential aspects from the problem using following assumptions:

- o Gas and liquid are considered as incompressible fluids
- o The flow has a symmetry in tangential direction and tangential component of liquid phase velocity is zero everywhere.
- o No mass transfer occurs between gas and liquid phases.

The sparging of gas in a stagnant pool of liquid generates complex, three dimensional turbulent flow. Therefore, it is essential to use an appropriate turbulence model to simulate this flow. Recently, Ranade [13] has reviewed the numerical simulations of dispersed gas-liquid flows. In the present work, we have used modified k-ε model proposed by Svendsen et al. [11]. To save space, the detailed transport equations are not included here. The terms which are different from those used by Svendsen et al. are described below.

Interphase Drag Force

The interphase drag force (acting on liquid phase, 1) is modelled as:

$$F_{D1j} = f_j \, (V_{2j}-V_{1j}) \qquad (1)$$

where f_j is the interphase friction factor between the gas and liquid in the general direction j; in that direction V_{2j} and V_{1j} represent the local velocities of the gas and liquid phase respectively. The factor f_j can then be related to drag coefficient C_D as:

$$f_j = (3/8)\rho_L \epsilon_G \epsilon_L C_D (V_{2j}-V_{1j})/R \qquad (2)$$

where R is effective radius of bubbles. The drag coefficient for the churn turbulent regime can be computed as (Ranade [8]):

$$C_D = [0.622/(.235+(1/Eo))] \; e_L^2 \qquad (3)$$

This equation was obtained for the axial flow of bubbles due to buoyancy forces. For the radial flow of bubbles due to radial pressure gradients may have a different drag coefficient than this. This aspect is discussed in the next section.

Simulation of Radial Profile of Axial Slip Velocity

Recent measurements have shown that there is a distinct profile of axial slip velocity of gas phase in bubble column reactors Yao et al. [2]. Preliminary numerical experiments of Ranade [8] also lend support to such a profile of axial slip velocity. Grienberger and Hofmann [10] have attempted to simulate this by introducing a radially varying drag coefficient which is difficult to justify. Instead one can simulate this by modelling the basic phenomena which causes this profile. Because of turbulent dispersion, bubbles get transported in the radial direction. When bubble gets transported radially outwards, because of the high velocity liquid wake attached to it, it retains the high velocity much longer than that it can retain in absence of the wake. Since in the two fluid model adapted here, there is no direct accounting for the wakes, we need to incorporate this effect by some means. In this work, we have done this by introducing an extra term in the source of the gas phase axial momentum equation. This term will be a function of wake characteristics, bubble dispersion coefficients, radial gas holdup and velocity profiles in the column. Assuming parabolic radial profiles for holdup and velocity and by accounting the other effects by an empirical constant, K_R this term can be written as:

$$F_R = K_R \, r^2 \qquad (4)$$

The empirical constant K_R can be adjusted to get reasonable predictions of the radial profile of axial slip velocity. Preliminary numerical experiments suggest the value of K_R as 2. A use of 3 fluid model (bubbles-bulk liquid-wake liquid) may eliminate this empiricism in future.

Simulation of Sparger Characteristics

In the detailed CFD based simulations of bubble column reactors, it is necessary to incorporate the effects of sparger characteristics on the distribution of the gas at the inlet. Since the pressure below the sparger, P_S is radially uniform and pressure just above the sparger is not uniform due to the radial holdup profile, the gas distribution at the sparger will not be uniform. The gas distribution at the sparger can be calculated using following relationship:

$$P_S - P(r,0) = k_V V_G(r) + k_T V_G(r)^2 \quad (5)$$

where P(r,0) is the pressure at the bottom of the column at radius r. k_V and k_T are proportionality constants. In the present simulation, pressure below the sparger, P_S was adjusted in such a way that the superficial gas velocity will remain at the predecided value. The gas distribution at the sparger is calculated at each iteration to account for the variation in P(r,0).

Boundary conditions used in the present simulation are shown in Figure 1. Wall functions have been used to impose a no-slip boundary condition at an impermeable wall. For simulating extra turbulence due to the gas bubbles, we have used a model proposed by Svendsen et al. [11] involving an exchange coefficient, C_b. Unless mentioned otherwise, the value of C_b was set to zero. The value of lift coefficient was set to -0.5 following Grienberger and Hofmann [10] unless mentioned otherwise. The sparging of gas at the bottom of the column was accounted via source terms in gas phase continuity equations for those computational cells which contain sparger holes at their bottom face. With this arrangement, various sparger characteristics can be accommodated conveniently into the model. It should be noted, however, that the model assumes that gas enters in the form of bubbles and therefore can not account for the gas jetting etc.

Solution of Model Equations

The set of transport equations were solved using finite volume method. All the results described below were obtained for 8x8 grids unless otherwise mentioned. The details of the numerical method and the computer code TPFLOW are given by Ranade [14]. The mean velocity, fractional hold-up and pressure need to underrelax to achieve convergence. The iterations for flow computations were continued till the mass residue becomes less than 1e-3.

RESULTS AND DISCUSSION

No attempt has been made to compare the predictions of our model with the experimental data since previous work of Grienberger and Hofmann [10] and Svendsen et al. [11] have demonstrated that reasonable agreement can be obtained by tuning appropriate model parameters. Instead, here emphasis is given on developing better understanding about major factors causing non-uniform radial distribution of gas holdup.

Influence of Radial Slip Velocity

In order to simplify the model and CFD code, Ranade [8] had assumed zero radial slip for their preliminary numerical investigation. In the present work, we have now incorporated a complete two fluid model in the TPFLOW code. The condition for zero radial slip can be imposed by setting the constant f_j in the interphase drag force formulation to very high value. The results obtained for the different values of f_j are shown in Figure 2. It can be seen that, imposing zero radial slip leads to incorrect results. Under such condition, gas holdup profiles remain almost flat in the upper portion of the column. This leads to negligible circulation in the upper portion of the taller columns as reported by Ranade [8]. All the following results were obtained by setting absolute slip velocity in the equation of f_r as 0.235 m/s. Typical profiles of gas holdup, turbulent viscosity and turbulent kinetic energy are shown in Figure 3.

Influence of Sparger Resistance

Recently Kolhatkar and Joshi [15] have demonstrated the importance of sparger resistance in the transition to the churn turbulent regime in bubble columns. Lower the resistance of sparger, this transition occurs earlier due to the increased tendency of nonuniform gas distribution at the sparger. Figure 4 shows gas distribution at the sparger for the three different sparger resistances at same superficial gas velocity. The liquid axial velocity profiles for these values are shown in Figure 5. It can be said that the nonuniform sparging also contribute to the gas holdup profile in the column in addition to the radial slip velocity. The present framework for accounting the sparger resistance is especially essential to simulate shorter bubble columns (H/D < 3) in which sparger can significantly affect the flow.

Influence of Lift Coefficient

Svendsen et al. [11] and Grienberger and Hofmann [10] have simulated the radial profile of gas holdup mainly through the use of negative lift coefficient. However, apart from this, the radial movement of bubbles due to the radial profile of hydrostatic head, non-uniform sparging at the inlet and an increase in axial slip velocity towards wall (due to the inertial effects of wake as discussed in section 3.2) also contribute for the nonuniformity in gas holdup. This can be seen from the results shown in Figure 6. Even if the lift coefficient is set to zero, there is non-uniform distribution of the gas leading to the circulation in the column. However, in this case two distinct circulation cells are developed unlike in the case of negative lift coefficient. This indicates that before fitting the value of lift coefficient to match the predictions with the experimental data, it is necessary to incorporate all the factors which contribute to the nonuniformity of the gas holdup as has been attempted in the present model.

Influence of Height to Diameter Ratio

The previous bubble column simulations of Svendsen et al. [11] and Grienberger and Hofmann [10] were restricted to only one height to diameter ratio (of 10). Figure 7 shows results obtained in the present work for height to diameter ratios of 2 and 8. It is interesting to note that the vector plots obtained for the H/D ratio of 2 show some internal circulation cells embedded in the overall circulation cell. Whereas such cells are not observed for the H/D ratio of 8 for the same number of grids (8x8). This must be because of very large aspect ratio of the computational cells. The computations of Svendsen et al. [11] and Grienberger and Hofmann [10] also did not show internal circulation cells presumably due to the same reason. Detailed examination of the results show that if the internal circulation cells are resolved, very low values of exchange coefficient, C_b (or even zero) give reasonable predictions of the turbulent kinetic energy. Where as Svendsen et al. [11] have used this value as high as 0.75 to get reasonable predictions. Extreme care need to be exercised in all the future CFD simulations of tall bubble columns to resolve the internal circulation cells with characteristic vertical dimension of the order of column diameter. This work is in progress now and it will hopefully lead towards better model for the extra turbulence generation by the gas bubbles and better understanding of the mechanism of the energy transfer from the gas phase to liquid phase.

CONCLUSIONS

The mathematical model for the two phase flow in bubble column has been developed. Numerical experiments were carried out to highlight the key variables in determining the flow characteristics of bubble columns. It has been shown that it is necessary to consider the radial slip velocity of bubbles in order to get the reasonable simulations. The influence of sparger characteristics were considered for the first time. It is necessary to incorporate all the factors which contribute to the nonuniformity of the gas holdup before fitting the value of negative lift coefficient. The effect of aspect ratio of the grids was highlighted to emphasis the resolution of the internal circulation cells. The model for simulating the extra turbulence generated by the bubbles can only be developed by properly ensuring the resolution of internal circulation cells.

ACKNOWLEDGEMENT

This research is supported by the Department of Science and Technology, Government of India (Sanction no. III 4(31)/90/ET).

NOTATION

C_b	Exchange coefficient as defined by Svendsen et al. [11]
C_D	Drag coefficient of bubble
D	Column diameter
Eo	Eotvos number
f_L	Lift coefficient
F_D	Interphase drag force
H	Height of the bubble column
k	Turbulent kinetic energy per unit mass
k_T, k_V	Sparger resistance constants
K_R	Empirical constant in Equation (4)
P	Pressure
P_S	Pressure below the sparger
r	Radial coordinate
R	Column radius
U	Radial mean velocity
W	Axial mean velocity
z	Axial coordinate

Greek symbols

ϵ	Phase holdup
ε	Turbulent energy dissipation rate per unit mass
ρ	Density

Subscripts

j	of j phase
G or 2	of gas phase
L or 1	of liquid phase

LITERATURE CITED

1. Devanathan, N., D. Moslemian and M.P. Dudukovik, Chem. Eng. Sci. **45**, 2285-2291 (1990).

2. Yao, B.P., C. Zheng, H.E. Gasche and H. Hofmann, Chem. Eng. Prog., **29**, 65-75 (1991).

3. Menzel, T., T. Weide, O. Staudacher, O. Wein and U. Onken, Ind. Eng. Chem. Res. **29**, 988-994 (1990).

4. Torvik, R., Investigation of 2- and 3-phase bubble columns, Ph. D. Thesis, Norges Tekniske Hogskole, Trondheim (1990).

5. Yu, Y.H. and S.D. Kim, Chem. Eng. Sci. **46**, 313-320 (1991).

6. Gross, R.W. and J.M. Kuhlman, Int. J. Multiphase Flow, **18**, 413-21 (1992).

7. Torvik, R. and H.F. Svendsen, Chem. Eng. Sci. **45**, 2325-2332 (1990).

8. Ranade, V.V., Chem. Eng. Sci, **47**, 1857-1869 (1992).

9. Neti, S. and O.E.E. Mohamed, Int. J. Heat and Fluid Flow, **11**, 204-

10. Grienberger, J. and H. Hofmann, Chem. Eng. Sci., **47**, 2215-2220 (1992).

11. Svendsen, H.F., H.A. Jakobsen and T. Torvik, Chem. Eng. Sci., **47**, 3297-3304 (1992).

12. Webb, C., F. Que and P.R. Senior, Chem. Eng. Sci., **47**, 3305-3312 (1992).

13. Ranade, V.V., Sadhana, **17**, 237-273 (1992).

14. Ranade, V.V., TPFLOW: A Codeting for Simulating Three Dimensional Two Phase Flows, Version II, NCL Internal Report (1992).

15. Kolhatkar, R.V. and J.B. Joshi, submitted for publication, Int. J. Multiphase Flow (1992).

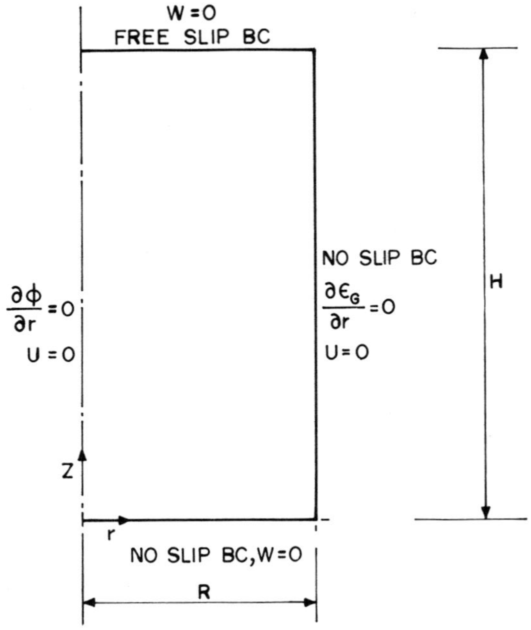

Figure 1: Solution domain

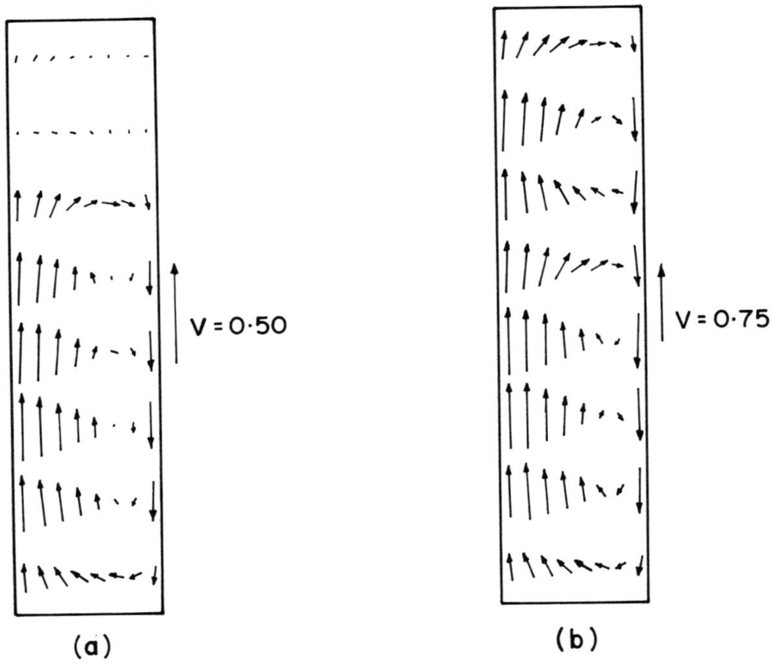

Figure 2: Influence of radial slip velocity

(a) no radial slip
(b) with radial slip

Column diameter, D = 0.3 m
Height to diameter ratio = 2
Lift coefficient, f_L = -0.5
Sparger resistance constant, k_T = 100

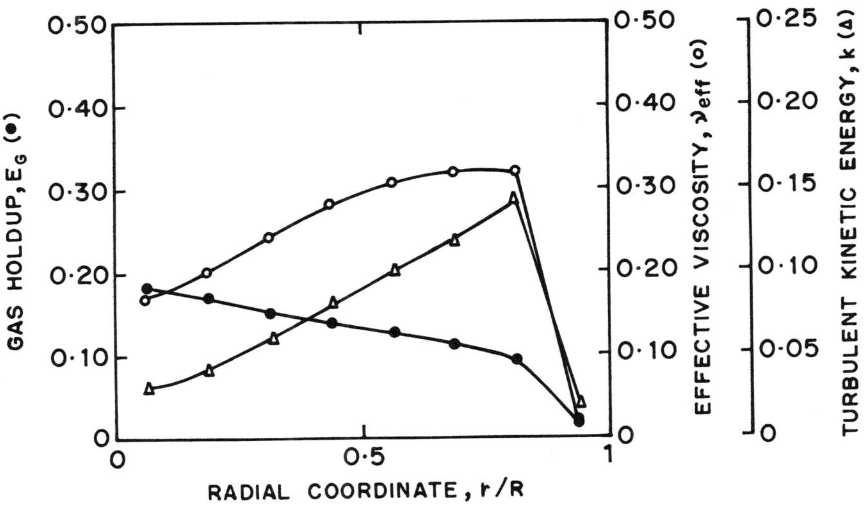

Figure 3: Typical profiles of gas hold-up, dimensionless turbulent viscosity and dimensionless turbulent kinetic energy

Other parameters are same as in Figure 2
With radial slip

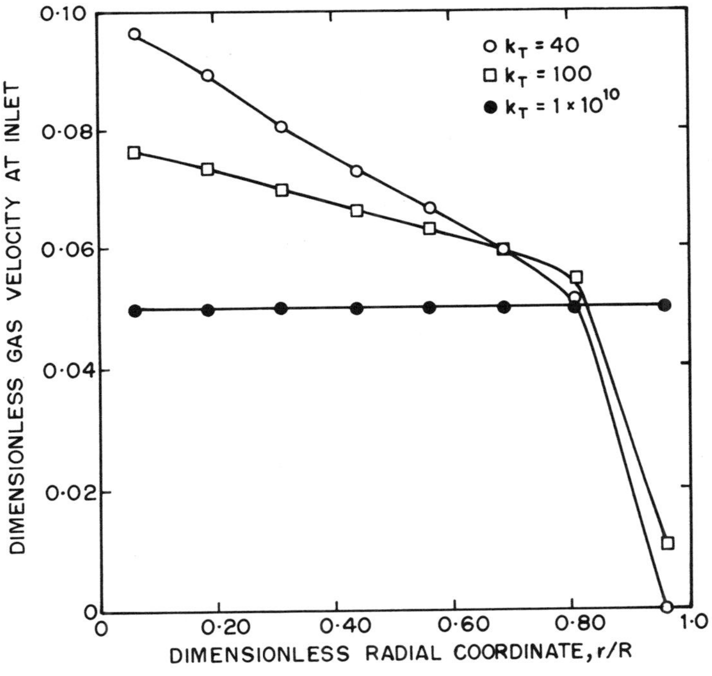

Figure 4: Influence of sparger resistance (k_T) on the gas distribution at the inlet

Other parameters are same as in Figure 3 except k_T

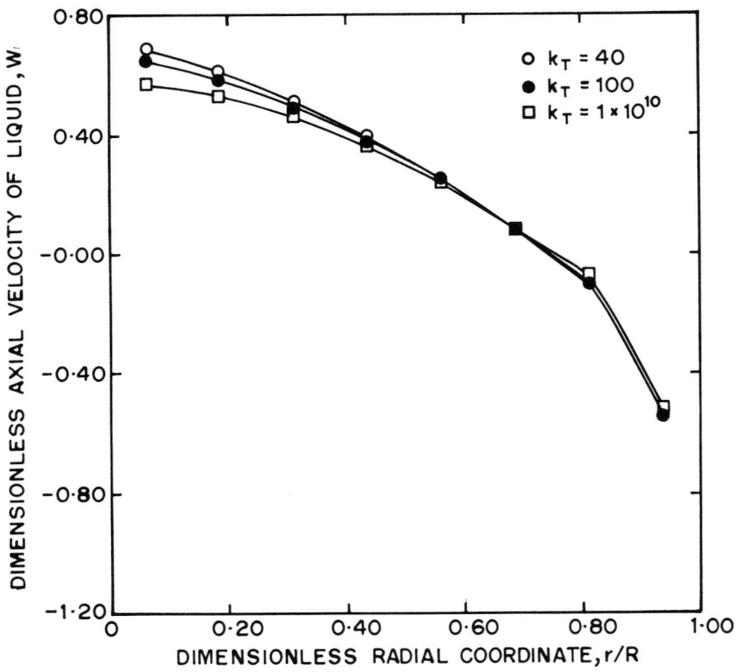

Figure 5: Influence of sparger resistance (k_T) on the axial velocity of liquid

Other parameters are same as in Figure 3 except k_T

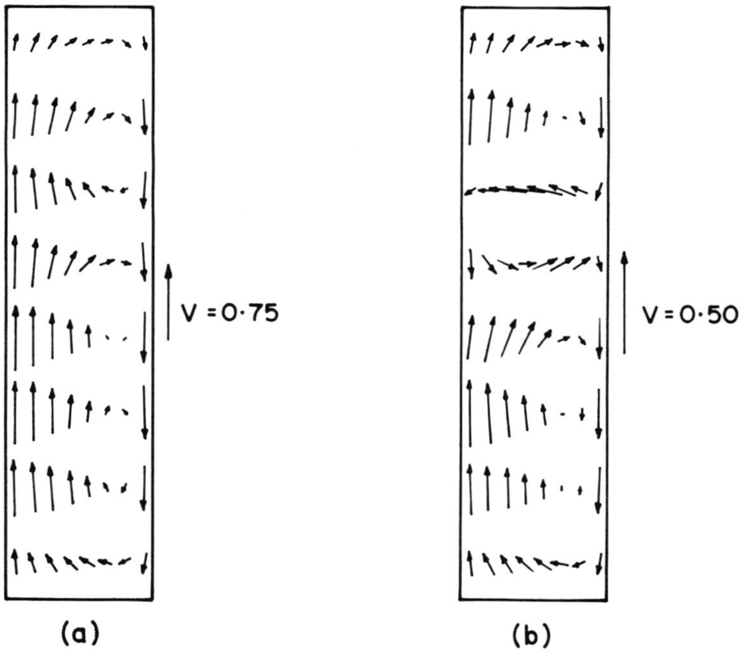

Figure 6: Influence of lift coefficient

(a) $f_L = -0.5$
(b) $f_L = 0.$

Other parameters are same as in Figure 3 except f_L.

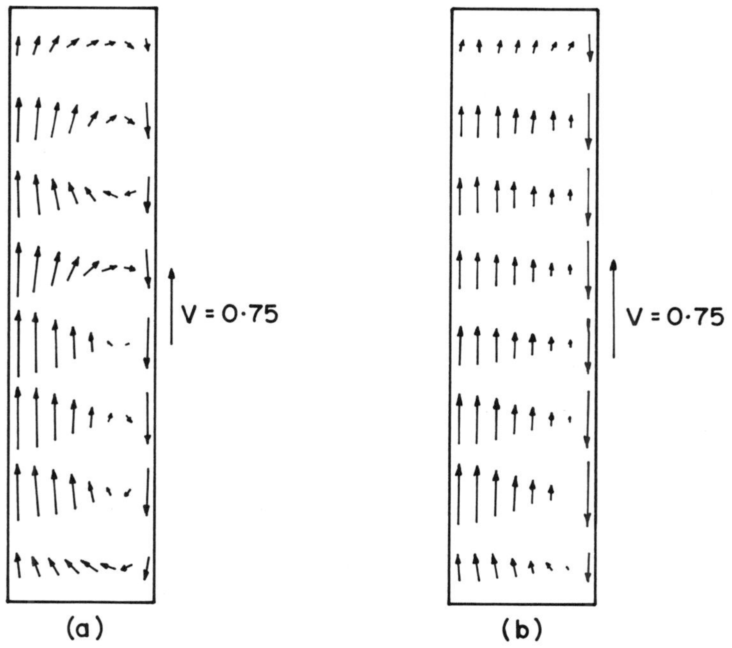

Figure 7: Influence of height to diameter ratio

(a) H/D = 2
(b) H/D = 8

Other parameters are same as in Figure 3 except H/D.

Effect of Mechanical Agitation on Alcoholic Fermentation

Enrique Galindo, Benito Román and Miguel Salvador

Depto. de Bioingenieria, Instituto de Biotecnologia,
Universidad Nacional Autonoma de Mexico, Apdo. Post. 510-3, Cuernavaca, Mor., 62271,
Mexico

The effect of mechanical agitation over alcoholic fermentation of sugarcane molasses by Saccharomyces cerevisiae *was evaluated in tanks of 0.002, 0.014, and 0.13 m^3 (nominal volume) equipped with Rushton turbines. Agitation slightly affected the yeast growth rate and sugar consumption rate, whereas the ethanol yield (kg of ethanol/kg of initial sugar) increased drastically as agitation was higher. When no agitation was used, the ethanol yield was about 64% of the theoretical. Agitation of about 0.4 kW m^{-3} achieved an alcohol yield of about 79% of the theoretical. Further increments in power supply (up to 1.6 kW m^{-3}) increased the yield to a maximum of 87% of the theoretical. When 1.6 kW m^{-3} were supplied to fermentors of 0.002, 0.014 and 0.13 m^3, the same ethanol yields were obtained, indicating that volumetric power input was a good scale-up criterion.*

Alcoholic fermentation is a process which is usually conducted in large tanks or "vats" of volumes up to 200 m^3, generally with no mechanical agitation. Strong gradients in environmental conditions are known to occur in those vessels. Mechanical agitation can be an alternative to decrease those gradients, potentially leading to improved fermentation performance.

Alcoholic fermentation is a widely studied process from the points of view of microbiology, kinetics and mode of operation of reactors. Nevertheless, very few works deal with the influence of hydrodynamics over fermentation performance and there are no publications covering the influence of mechanical agitation over the alcoholic fermentation. Valdeolivas *et al* (1) have published data about the influence of the reactor geometry (0.002 and 0.004 m^3 in volume) over the evolution of alcoholic fermentation. Renger & Luyben (2) have shown the advantages of using a conical tank in beer production.

Very little has been published as well regarding the effect of mixing in other anaerobic processes, such as the acetone-butanol fermentation (3,4).

Ethanol producers, especially fuel ethanol manufacturers, should run their processes at the highest possible yields in order to be economically competitive. Many fuel alcohol plants conduct fermentation in tanks equipped with mechanical agitation; however, there are no reports in the open literature about the potential advantages of using mechanical agitation in this anaerobic fermentation.

The aim of this work was to evaluate the effect of agitation speed (*i.e.* power supply) in laboratory reactors as well as in bench and pilot fermentors upon the performance of alcoholic fermentation using *S. cerevisiae* and sugarcane molasses as substrate.

MATERIALS AND METHODS

Strain and molasses. The yeast *Saccharomyces cerevisiae* (active dry yeast, Safmex, México) was used in all experiments. A batch of sugarcane molasses was provided by a local sugar mill. Molasses were stored at room temperature. The same batch was used for all experiments.

Analytical techniques. Biomass was estimated by optical density at 460 nm in a spectrophotometer (Beckman, model 35), using a blank with uninoculated molasses broth. Optical density was correlated with dry weight by a standard curve. Ethanol was measured by gas chromatography in a Varian (model 3700) equipment. Broth samples were subjected to extraction with n-butanol and the organic phase was injected to the chromatograph. Sugar content was measured (as glucose) by an enzyme analyzer (YSI, model 27) as described elsewhere (5).

Fermentation broth. The broth contained molasses sufficient to prepare a dilution of 26° Bx, and (in kg m^{-3}): ammonium sulphate, 4; ammonium phosphate, 0.05; ammonium bifluoride, 0.125 and penicillin V, 0.002. Distilled water was used in all cases. All medium components were industrial grade. The broth sugar content was adjusted to have 170 kg m^{-3} (measured by the enzyme analyzer) by addition (when necessary) of small quantities of molasses. The composition of the medium used for the development of inocula was as follows: molasses, 17° Bx and (in kg m^{-3}): ammonium sulphate, 8; ammonium phosphate, 0.005; ammonium bifluoride, 0.125; penicillin V, 0.002. The inoculum was grown for 12 h in a shaker (or bench fermentor) at 30°C. Initial yeast concentration was adjusted to 2 kg m^{-3}.

Fermentors. Tanks of 0.002, 0.014 and 0.13 m^3 (nominal volume) equipped with three Rushton turbines were used. The dimensions of the tanks and of the impellers are given in table 1. Power consumption in the 0.002 and 0.014 m^3 tanks was evaluated by an air bearing dynamometer (6) on which the tanks were mounted. Power consumption in the 0.13 m^3 tank was calculated from literature (7) correlations.

Table 1. Geometric dimensions of the fermentors

	Fermentor volume (m^3)		
	0.002	0.014	0.13
T	0.11	0.21	0.416
D/T	0.44	0.33	0.33
W_1/D	0.19	0.21	0.19
L_1/D	0.25	0.25	0.28
N_b	4.00	4.00	4.00
W_b/T	0.11	0.11	0.09
Z/D	1.5	1.5	1.5

Anaerobic conditions. Anaerobiosis was assured bubbling nitrogen into the broth until the dissolved oxygen tension (measured with an Ingold electrode) was virtually zero. To avoid surface aeration, the fermentors were filled to up to 95 % of their nominal volume. Oxygen tension was measured throughout the fermentation runs and was virtually zero.

RESULTS AND DISCUSSION

Fermentations without mechanical agitation

A typical kinetic evolution of a fermentation process without mechanical agitation is shown in figure 1. As shown in table 2, similar final concentrations of biomass and ethanol were found in the different scales tested. However, fermentation time was longer as the scale increased. These differences may be the result of a scale effect and/or the heterogenity of the system. As the scale increased, the circulation times (due to bubble rising) are longer and hydrostatic pressure is higher. Thus, if the CO_2 generation rate is similar, the residence time of CO_2 is a function of the liquid height and gradients of CO_2 will be formed. This would be valid for biomass, glucose and ethanol concentrations as well. It is important to notice that the ethanol yield ($Y_{P/S}$) at different scales represented only 64 % of the maximum theoretical yield (0.51). The relatively low yields obtained indicated that the process conditions are susceptible to improvements.

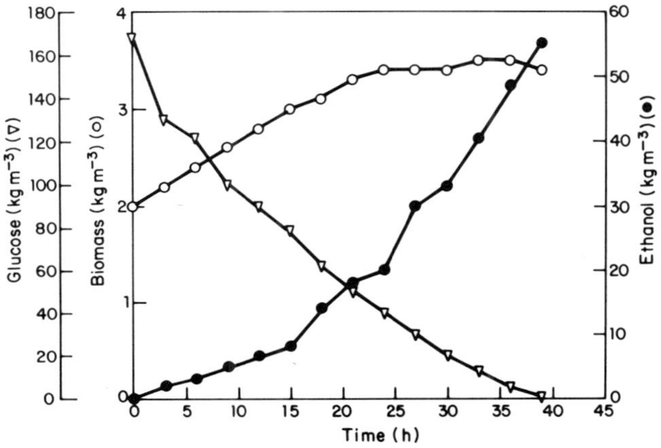

Figure 1. Evolution of alcoholic fermentation in a 0.014 m^3 fermentor with no mechanical agitation.

Table 2. Final results of alcoholic fermentation with no mechanical agitation

	Fermentor volume (m^3)		
	0.002	0.014	0.13
Biomass (kg m^{-3})	3.4	3.4	3.4
Ethanol (kg m^{-3})	55.7	55.1	55.0
Fermentation time (h)	35.0	39.0	40.0
Yield			
$kg_{ethanol}$ /kg_{sugar}	0.328	0.324	0.323

Fermentations with mechanical agitation

In order to evaluate the effect of agitation on the fermentation process, different agitation speeds were tested. Figure 2 shows an example of the effect of mechanical agitation over alcoholic fermentation kinetics, of cultures developed in a 0.014 m^3 fermentor. Slightly higher substrate consumption rates were observed as agitation increased. Biomass profiles were about the same. The most dramatic effect of agitation was on the evolution of ethanol, which showed an important increase after 25 h of fermentation. Compared with the non-agitated process, the maximum ethanol concentration was 36 % higher and fermentation time was 4 h shorter. The $Y_{p/s}$ reached up to 87 % of the theoretical yield.

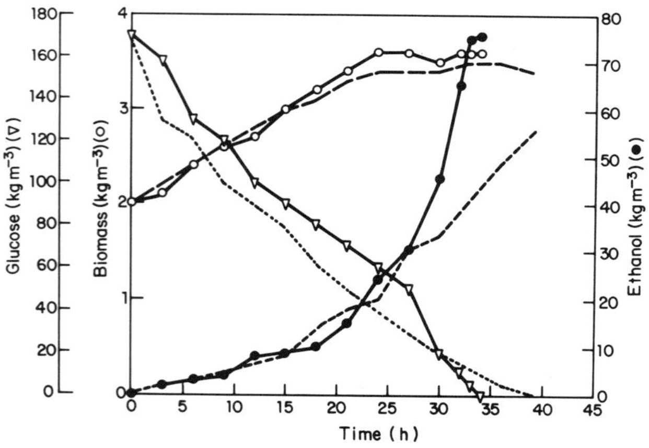

Figure 2. Evolution of alcoholic fermentation in a 0.014 m^3 fermentor stirred at 7.16 s^{-1} (1.62 kW m^{-3}). Lines without data points show the results with no mechanical agitation. (...) glucose, (---), ethanol, (– –) biomass.

Figure 3 shows the relationship between the volumetric power drawn (P/V) and the ethanol yield, normalized with respect to the non-agitated conditions (static). A pronounced improvement in fermentation yield was observed as the power drawn was higher (up to about 1.6 kW m^{-3}). Under relatively mild agitation conditions (0.2-0.4 kW m^{-3}) the yield was up to 23 % higher if compared with the non-agitated process. Power drawn higher than 1.6 kW m^{-3} did not improve the yield, which was 87 % the theoretical maximum.

The influence of mechanical agitation over the ethanol yield can be due to various reasons; including better yeast suspension, improved heat and CO_2 removal and minimizing physicochemical gradients.

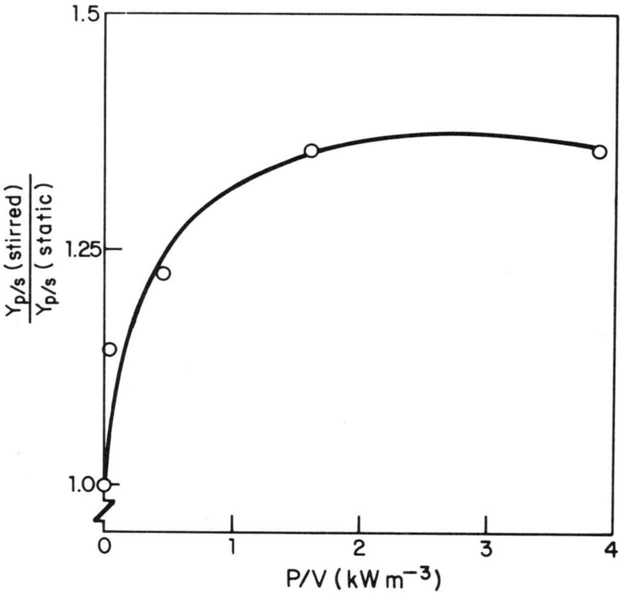

Figure 3. Normalized ethanol yield as a function of power drawn. 0.002 m^3 fermentor with 3 Rushton turbines.

Scale up

It was decided to use P/V as the scaling-up criterion. It is well known that volumetric power input is proportional to the cube of stirring speed and to the square of impeller diameter, *i.e.*:

$$\frac{P}{V} = \alpha\, N^3 D^2 \qquad (1)$$

For the same P/V values under different scales, it follows that:

$$\frac{N_1^3}{N_2^3} = \frac{D_2^2}{D_1^2} \qquad (2)$$

Using equation (2) we calculated the N values for scaling-up the process to 0.014 and 0.13 m^3, keeping P/V constant at 1.62 kW m^{-3}. The resulting N values where 7.16 and 5.0 s^{-1} for the 0.014 and 0.13 m^3 fermentors, respectively. The results shown in table 3 indicated that P/V was a good criterion for scaling-up ethanol production process since the fermentation was reproducible at the different scales tested.

Table 3. Scale-up of alcoholic fermentation maintaining P/V constant at 1.62 kW m^{-3}. Results at the end of fermentations.

	Fermentor volume (m^3)		
	0.002	0.014	0.13
N (s^{-1})	10.0	7.16	5.0
Biomass (kg m^{-3})	3.6	3.6	3.6
Ethanol (kg m^{-3})	75.5	75.1	75.0
Fermentation time (h)	32.0	34.0	35.0
Yield $kg_{ethanol}/kg_{sugar}$	0.44	0.44	0.44

CONCLUSIONS

Ethanol production is affected by mixing conditions. Power drawn strongly affect fermentation yield up to a value of about 1.6 kW m^{-3}, where ethanol yields were 87 % of the theoretical maximum. Fermentation results were succesfully scaled-up in fermentors of 0.014 and 0.13 m^3 using P/V as criterion. Mechanical agitation can improve industrial alcoholic fermentation performance and there fore its study deserves further work in larger scales. Such studies should also take into account economical considerations.

NOMENCLATURE

D	Impeller diameter (m)
L_i	Impeller blade lenght (m)
N	Impeller speed (s^{-1})
N_b	Number of baffles (-)
P	Power drawn (W)
T	Tank diameter (m)
V	Volume (m^3)
W_b	Baffle width (m)
W_i	Impeller blade width (m)
$Y_{p/s}$	Ethanol yield on substrate ($kg_{Et} kg_{sug}^{-1}$)
Z	Distance between impellers (m)

Subscripts

1	Laboratory scale (0.002 m^3)
2	Bench or pilot scale (0.014 and 0.13 m^3)

LITERATURE CITED

1. Valdeolivas, G.E.I., M.L.R. Vairo. and W. Borzani. *Arq. Biol. Tecnol.* (Brasil), **24**, 3, 361 (1981).

2. Renger, R.S. and K.Ch.A.M. Luyben. Fluid dynamics of beer fermentation, In: R. King (Ed), *Proc. 2nd Int. Conf. Bioreactor Fluid Dynamics*, Cambridge, BHRA The Fluid Engineering Centre, Cranfield, Bedford, p. 277 (1981).

3. Doremus, M.G., J.C. Lindon. and A., Moreira. *Biotechnol. Bioeng*, **27**, 852 (1985).

4. Yerushalmi, L. and B. Volesky, *Biotechnol. Bioeng.*, **27**, 1297 (1985).

5. García, J.L., M. García-Garibay, M. Salvador and E. Galindo. Limitations on the use of an enzymatic electrode for the measurement of sugars in molasses-based alcoholic fermentation broths, *Biotechnol. Techniques*, submitted (1992).

6. Reséndiz, R., A. Martínez, G. Azcanio and E. Galindo. *Chem. Eng. Technol.*, **14**, 105 (1991).

7. Ulbrecht, J.J. and G.K. Patterson. Mixing of liquids by mechanical agitation, Gordon & Breach Science, Pub., New York, (1985).

Gas Dispersion at High Aeration Rates in Low to Moderately Viscous Newtonian Liquids

M. E. Sensel and K. J. Myers

Chem. and Materials Eng. Dept., University of Dayton, Dayton, OH 45469-0246

J. B. Fasano

Chemineer Inc., 5870 Poe Ave., Dayton OH 45414

Data is provided for the effects both of high gassing rate and moderate viscosities on flooding and gassed power demand for the D-6 and CD-6 impellers. Results for aeration numbers to 1.2 in an air-water system are compared to literature data available only for low gas rates. It is shown that predictive algorithms for low gas rates cannot be extended to high gas rates without significant error. Higher viscosities were observed to require higher power to prevent flooding and an increase in gassed power.

Introduction

In today's industrial marketplace, there is an ever increasing need for gas dispersion design information at high gassing rates. Much of the literature has generally limited aeration numbers, N_a, to 0.2, while industrial applications are quite commonly above 0.3, and some applications are in the neighborhood of 0.7. Aeration numbers reported in this study ranged from zero to 1.5.

Existing literature has also centered nearly all of its investigation on air-water systems. More often than not, however, industrial applications require a more viscous continuous phase, and within this context there is very little information available for the designer. Some work has been done on severely non-Newtonian systems, such as xantham gum. However, the authors believed that a clear understanding of how Newtonian systems behaved was essential before examining non-Newtonian systems. Effects of viscosity up to 350 cp (mPas) were examined in this investigation.

Experimental System

The experimental apparatus was designed to model actual industrial process configurations. The majority of the tests were run in a 15.75" (0.400m) inner diameter, 2:1 elliptical dish bottomed tank. Additional tanks were used to determine the scale up effect of the flooding phenomena and were 11.5" (.292m), 19.5" (0.495m), and 60" (1.52m) diameter. Although the following description will refer only to the 15.75" (0.400m) tank, all tanks used for these gas dispersion studies were geometrically identical.

The 15.75"(0.400m) tank was constructed of a transparent plexiglass so that flow patterns could easily be observed. The unaerated liquid height was 70 percent of the tank diameter, or a volume of 9.96 gallons (0.0377m^3). The tank was fitted with four standard, stainless steel baffles, 1.5" (0.0381m) in width. The baffles were placed 0.375" (0.00952m) from the tank wall, extending the entire length of the straight side of the tank.

Two 6-blade, disk impellers were studied, a Chemineer D-6, impeller, and a Chemineer CD-6 impeller. Each impeller was 5.7" (0.145m) diameter, which corresponds to an impeller diameter to tank diameter ratio, D/T, of 0.36. Each impeller had a blade width to diameter ratio, W/D, of 0.2. The schematics of both impellers can be found in Figure 1. Both impellers were placed at an off-bottom clearance of one-third of the liquid depth, or 3.667" (0.0931m).

A sparge ring was used to evenly distribute the gas supplied to the system. The sparge ring consisted of a 0.5" (0.0127m) diameter tube formed into a 4" (0.102m) diameter ring with 10 equally spaced holes, 0.125" (0.00318m) diameter. The sparge ring was centered in the tank by a bracket attached to the baffles, 2.105" (0.0535m) from the bottom of the tank.

The experimental drive used for the experiments had a 2 hp (1.34kW) motor with a range of speed up to 2000 rpm (33.3 rps). The speed was measured using a zero velocity magnetic pickup. A calibrated strain gauge torque cell was used to determine torque requirements. The speed and torque indicated were electronically combined to display impeller power. All three were displayed on a digital indicator.

The transition between impeller flooding and loading was detected both by visual observation of the bulk flow and a drop in power demand. All flooding data was taken as the impeller went from flooded to loaded. All holdup and power draw data was taken as the impeller went from loaded to flooded. Holdup was determined from a measurement made by visual observation.

Flooding

Various investigators have approached descriptive techniques for flooding in different ways. Warmoeskerken and Smith[1] studied flooding of a D-6 turbine in air-water. They proposed that there is a linear relationship between the aeration number and the Froude number which is independent of scale.

$$N_a = a * N_{Fr} \qquad (1)$$

In their study the tank diameter to impeller diameter ratio was held constant at 2.5 over a range of tank diameters. The results of this study in air-water were plotted using this suggested relationship over a similar range of Froude numbers, Figure 2. This study was performed with impellers having a tank diameter to impeller diameter ratio (T/D) of 2.76. Warmoeskerken and Smith found this proportionality constant to be 1.2. The data of this study in the lower range of aeration numbers justifies the linear relationship suggested by Warmoeskerken and Smith. However, it can be observed from Figure 2 that linearity only exists for aeration numbers of 0.2 or lower.

Further, scale effects are observed, Figure 3. Note that, for a given Froude number, as scale increases, impellers will flood at higher aeration numbers.

Nienow et al.[2] report a different relationship to predict the flooding transition. Over a limited range they showed that the flooding transition is a function of the aeration number and ND, the product of the impeller speed and the impeller diameter.

$$N_a = \alpha * N * D \qquad (2)$$

When the data of this study are plotted, the results are also linear at low values of ND as seen in Figure 4, and as suggested by Nienow et al. However, above ND values of 1.5 ft/sec (0.457m/sec) this linear relation no longer holds. Figure 5 is a similar plot of the results from this study for various scales. A scale effect is obvious, and the aeration number-ND relation at flooding is not linear across the entire range investigated in this study.

Nienow et al. proposed that there is a scale effect associated with flooding. They proposed the relationship,

$$N_a = f(N * D * T^{-0.4}) \qquad (3)$$

which also collapses this study's results for the CD-6 impeller, Figure 6. The fitted equation for this study's results is

$$N_a = 0.0675 * N * D * T^{-0.4} \qquad NDT^{-0.4} \leq 1.6$$

$$N_a = 0.0231 * (N * D * T^{-0.4})^{3.75} \qquad NDT^{-0.4} \geq 1.6 \qquad (4)$$

$$\text{where } N = \sec^{-1}, \quad D = ft, \quad T = ft$$

The average error of the fitted correlation is 8.28 %.

Thus far data has been presented for air-water systems. In addition, flooding transition studies were also performed in viscous, Newtonian solutions. Similar to water systems, the D-6 impeller floods before the CD-6 impeller for all viscosities studied. Figures 8 and 9 are examples of the results of these comparisons. Viscosity does not have a large influence on the flooding behavior of the CD-6 impeller. Initially the data falls on nearly the same line. At ND values above 2.5 ft/sec (0.762m/sec) though, the data separates slightly for the CD-6 impeller but not nearly to the extent that the D-6 impeller data separates. The rise in viscosity slightly decreases the gas handling capability of the CD-6 impeller but rather substantially decreases the gas handling capability of the D-6 impeller. As ND increases, the difference in gas handling ability between the D-6 and CD-6 impellers becomes larger.

Power Draw

It is important in designing agitators for gas dispersion applications to quantify the unaerated power demand at low Reynolds numbers since there is a growing number of high viscosity gas dispersion applications. To characterize the power demand, the impeller power number, N_p,

$$N_p = \frac{P_o}{N^3 * D^5 * \rho} \qquad (5)$$

is related to the impeller Reynolds number, N_{Re},

$$N_{Re} = \frac{N * D^2 * \rho}{\mu} \qquad (6)$$

graphically. Dickey and Fenic[3] illustrated the power number dependence of various impellers, including the D-6 impeller, on the Reynolds number. The power number dependence of the D-6 impeller on the Reynolds number was repeated in this study, Figure 9, and compares well with literature data. All power number experiments were performed in a flat-bottom tank with a impeller diameter to tank diameter ratio of 0.323. Only Newtonian fluids were studied. The D-6 impeller power number exhibits a minimum at a Reynolds number of about 200, and this minimum power number is approximately two-thirds of the fully turbulent power number. The D-6 power number increases dramatically as Reynolds numbers decrease below 100. The dip in the curve causes a difficulty in properly loading a D-6 impeller whose Reynolds number may change throughout the batch or from run to run.

The experiments were repeated with the CD-6, curved blade impeller. The results are shown in Figure 10. For Reynolds numbers under 200, the CD-6 power number behavior resembles that of the D-6 impeller. Unlike the D-6 impeller, the CD-6 does not exhibit an intermediate minimum power number. Instead, the CD-6 power number is relatively constant above Reynolds numbers of 200, and the fully turbulent, CD-6 power number is approximately two-thirds that of the D-6 impeller. The power required for the CD-6 impeller and therefore the motor capacity required, is less than the D-6 impeller under similar conditions. Without the intermediate Reynolds number range minimum, proper sizing is more easily accomplished when Reynolds numbers are always expected to be above 100.

When gas is added to an agitated system, the power demand almost invariably decreases. This is partly due to the decreased density of the solution and partly to the reduction in drag caused by gas cavities that form behind the impeller blades.

The ratio of the aerated power draw to the unaerated power draw is an effective way to quantify the drop in the power under gassed conditions. Figure 11 is an example of the power demand ratio of the D-6 impeller. Initially increasing the aeration number decreases the power demand of the impeller drastically. This influence is approximately constant at aeration numbers above 0.20. The power demand also decreases as the Froude number increases, and one can observe that the influence of the Froude number decreases at higher Froude numbers.

A graphical representation of the power demand of the CD-6 is provided in Figure 12. The power drop of the CD-6 is less than that of the D-6 impeller, and Figure 13 compares the power draw ratio of both impellers at a constant Froude number. The drop in power of the CD-6 is significantly less than that of the D-6 impeller for aeration numbers above 0.1.

Figures 14 and 15 illustrate the effect of viscosity on the power drop of the D-6 impeller and CD-6 impellers, respectively, at a fixed Froude number. Increasing viscosity reduces the power drop of both impellers. However, viscosity has less of an influence on the performance of the CD-6 impeller than the D-6 impeller. A empirical correlation that directly incorporates the effect of viscosity is

$$\frac{P_g}{P_o} = 1 - (a\mu + b)\tanh(cN_a) * N_{Fr}^d \qquad (7)$$

The coefficients and exponent from Equation 7 for both the D-6 impeller and the CD-6 impeller can be found in Table 1. Both impellers exhibit some scatter, and the average error of both correlated equations is less than ten percent.

Table 1: Coefficients and Exponent for Equation 7

	a	b	c	d	Avg. % Error
D-6	-0.000715	0.723	24.54	0.25	7.09
CD-6	-0.000115	0.440	12.077	0.37	5.20

Holdup

The fractional gas holdup in different viscosity solutions was determined at various combinations of impeller speeds and volumetric gas flows. The holdup definition used for this study is given in Equation 8.

$$\varepsilon = \frac{V_G}{V_L + V_G} \tag{8}$$

Figure 16 is an example of the data taken with a D-6 impeller in air-water. The holdup increases as either the aeration number, $Q/(ND^3)$, or the Froude number, N^2D/g, increase. Figure 17 illustrates the results of the CD-6 under similar conditions. The difference in the impeller type, D-6 impeller or CD-6 impeller, seems to have no significant effect on the holdup in air-water systems, as Figure 18 shows. It should be noted, however, that although the holdups are quite similar, the power investment required of each impeller can be significantly different.

Figures 19 and 20 illustrate how at higher Froude numbers, increasing the viscosity decreases the fractional gas holdup. At lower Froude numbers the fractional gas holdup is relatively unaffected by viscosity. An explanation for the effect of viscosity on gas holdup is that increasing liquid viscosity leads to lower bubble rise velocities. At higher aeration numbers, the bubbles issuing from the sparger coalesce with the slow-moving bubbles in the liquid. This coalescence leads to the formation of large gas bubbles that rapidly rise through the liquid, thereby reducing the gas holdup. The influence of viscosity on gas holdup appears to be approximately constant above viscosities of 100 centipoise.

Over the years many studies have been performed on fractional holdup. Many of these have been correlated in the form

$$\varepsilon \propto \left(\frac{P_g}{V_L}\right)^a * (U_g)^b \tag{9}$$

This correlation form is good for estimates of holdup, but lacks high accuracy. One reason could be that this form may fail to account for the holdup that occurs from gas flow without agitation. The holdup from gas flow can be integrated into the power per volume term if the power dissipated by the gas flow is measured. Another way to account for this would be to measure the fractional holdup from gas flow without any agitation. This can then be added to the agitated holdup as shown in Equation 10.

$$\varepsilon = \varepsilon_{gas} + \varepsilon_{agitation} \tag{10}$$

In this study the fractional holdup was correlated to the following form:

$$\varepsilon = aU_g^b + c\left(\frac{P_g}{V_L}\right)^d * (U_g)^e \tag{11}$$

The coefficient a and exponent b were determined by measuring unagitated holdup as a function of superficial gas velocity. The coefficient c and exponents d and e were then determined from agitated fractional holdup, gas power per volume and superficial gas velocity data while a and b were held constant. These coefficients and exponents are shown in Tables 2 & 3 for various viscosities. The correlation average percent error is also included in these tables. The parameters do not exhibit any apparent trend. The changing bubble rise rate caused by increasing viscosity as described above could be an explanation for the lack of trend.

Smith[4] suggests a different correlating equation for predicting fractional holdup, Equation 12.

$$\varepsilon = a * (N_{Re}\, N_{Fr}\, N_A)^{0.35} \tag{12}$$

He believes that this equation is

Table 2: Coefficients and Exponents for equation 10 (D-6 Impeller)

μ	a	b	c	d	e	Avg % Error
1	0.075	0.524	0.014	0.907	0.753	23.57
8	0.331	1.070	0.014	0.783	0.512	12.57
100	0.094	0.579	0.032	0.398	0.524	16.91
200	0.062	0.453	0.034	0.467	0.662	10.57

Table 3: Coefficients and Exponents for equation 10 (CD-6 Impeller)

μ	a	b	c	d	e	Avg % Error
1	0.075	0.524	0.031	0.778	0.851	13.65
8	0.331	1.070	0.043	0.508	0.524	10.98
100	0.094	0.579	0.081	0.245	0.708	8.78
200	0.062	0.453	0.058	0.322	0.652	12.30

independent of scale. This holdup correlation was used to analyze the results of this study. Smith originally intended for this correlation to be applied to only air-water systems, and Figure 21 illustrates that indeed this method does not adequately describe the effect of liquid viscosity on gas holdup. Smith's correlation could be adapted to describe viscosity effects by allowing the constant, a, to be a function of viscosity. Rather than follow this approach, the results of this study were fitted in the following manner which is similar in form to Smith's suggested correlation.

$$\varepsilon = 10.5 * (N_a * N_{Fr})^{0.5} * N_{Re}^{0.1} \quad (13)$$

This correlation collapses the data taken at various viscosities into a single line, shown in Figure 22. Comparison of the correlations of Equation 12 and 13 indicates that viscosity does not influence the fractional holdup to the extent implied by Smith's correlation. The correlation of Equation 13 has an average error of approximately 25%. However, at fractional gas holdups above 5%, as typically encountered in industrial applications, the average error is only 9%.

Summary and Conclusions

The often observed linear effects reported by numerous investigators for flooding, when extended to high aeration numbers, does not hold and care in design must be exercised. In addition, viscous effects can reduce the maximum amount of gas an impeller can handle. However, the extent to which the maximum gassing rate can be reduced is dependent upon the style of impeller. Scale effects also may not be ignored. For a given ND parameter, a larger tank will flood at a lower aeration number.

It is important in describing an impeller's gassed power via a P_g/P_o versus aeration number plot to include Froude number as a parameter. Low Froude number applications will always tend to unload less. Impeller style can drastically affect the gassed power characteristics of an impeller, with the newer style concave blade type of impeller, CD-6, exhibiting much less unloading under gassed conditions. The designer must also be concerned with viscosity. Higher viscosity applications will tend to draw more power even though the impeller Reynolds number is well with the turbulent range. If this is not properly taken into account, an agitator may overload during operation.

Holdup has historically been difficult to model. However, Equation 13 should provide a fairly accurate estimate for the gassing of Newtonian fluids from 1 to 350 cp (mPas).

Nomenclature

D, impeller diameter, in. *

N, impeller speed, rpm. *

N_a, aeration number, Q/ND^3, dimensionless

N_{Fr}, Froude number, N^2D/g, dimensionless

N_{Re}, Reynolds number, $\rho ND^2/\mu$, dimensionless

N_p, power number, $P_o/\rho N^3D^5$, dimensionless

P_g, aerated power demand, hp. *

P_g/V, aerated power per volume, hp/1000 gal *

P_o, unaerated power demand, hp *

Q, volumetric gas flow, scfm *

T, tank diameter, in. *

U_g, superficial gas velocity, ft/sec *

V_G, volume of gas, ft^3 *

V_L, volume of liquid, ft^3 *

ε, fractional gas holdup, dimensionless

ρ, density, g/ml *

μ, viscosity, cp *

* units unless indicated otherwise

References

1. Warmoeskerken, M. M. C. G., and Smith, J. M., "Description of the Power Curves of Turbine Stirred Gas-Liquid Dispersions", 4th European Conference on Mixing, Noordwijkerhout, Netherlands, April 1982.

2. Nienow, A. W.; Wisdom, D. J.; and Middleton, J. C., "The Effect of Scale and Geometry on Flooding, Recirculation, and Power in Gassed Stirred Vessels", 2nd European Conference on Mixing, Cambridge, England, 1977.

3. Dickey, D. S., and Fenic, J. G., "Dimensional Analysis for Fluid Agitation Systems", Chemical Engineering, Jan. 5, 1976, p. 139.

4. Smith, J. M., "Simple Performance Correlations for Agitated Vessels", 7th European Conference on Mixing, Brugge, Belgium, p. 223, 1991.

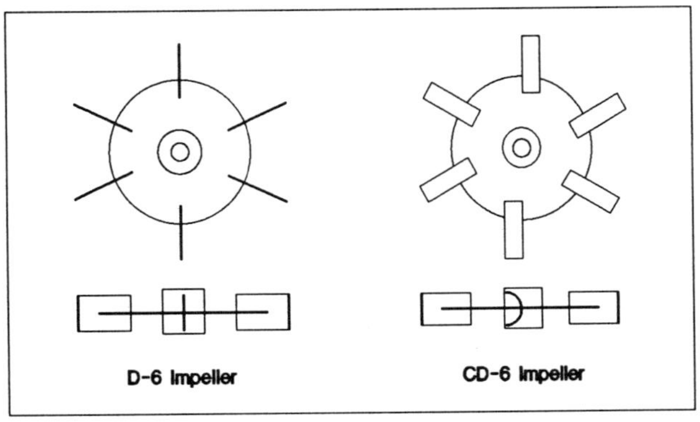

Figure 1

Flooding Point in an Air-Water System

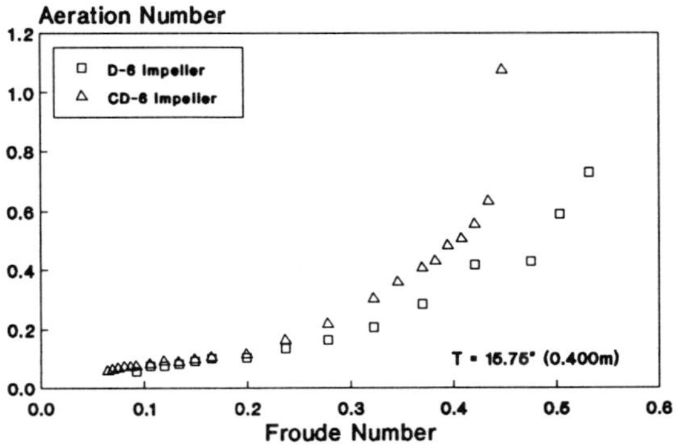

Figure 2

Scale Effects on Flooding Behavior

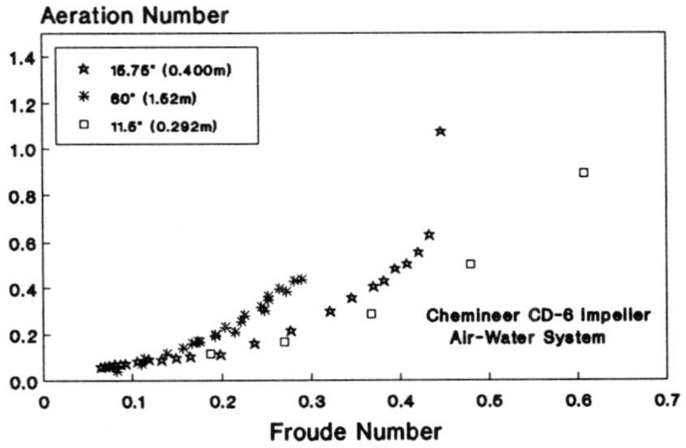

Figure 3

Flooding Point in an Air-Water System

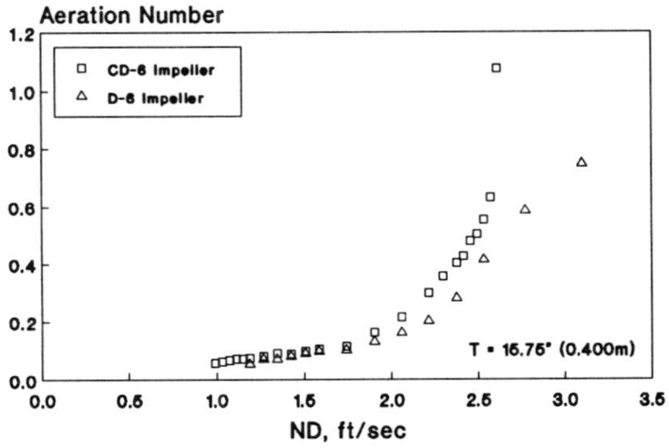

Figure 4

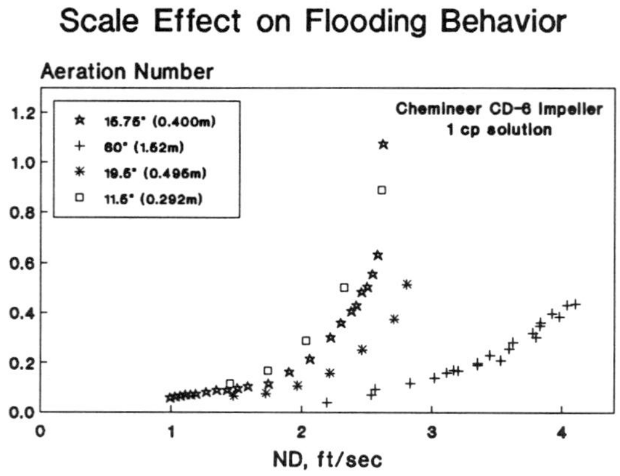

Figure 5

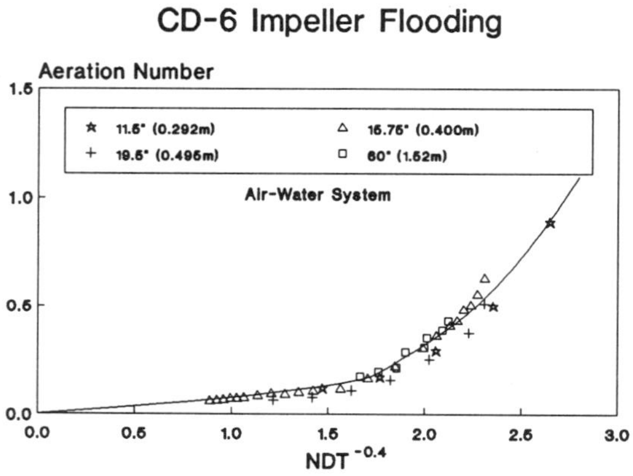

Figure 6

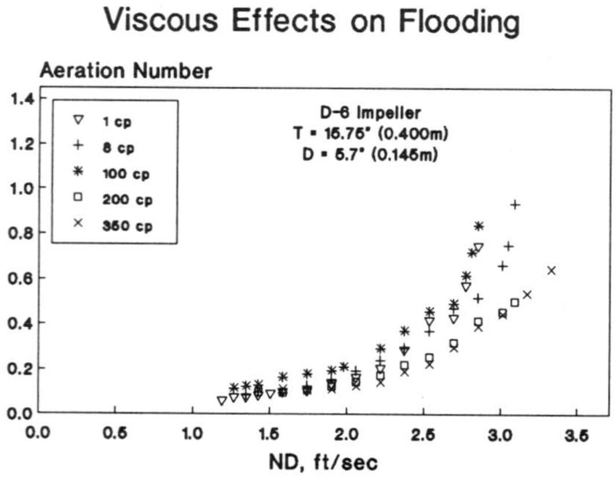

Figure 7

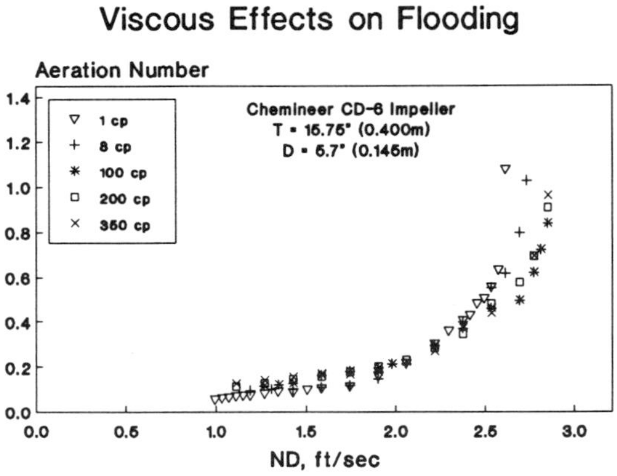

Figure 8

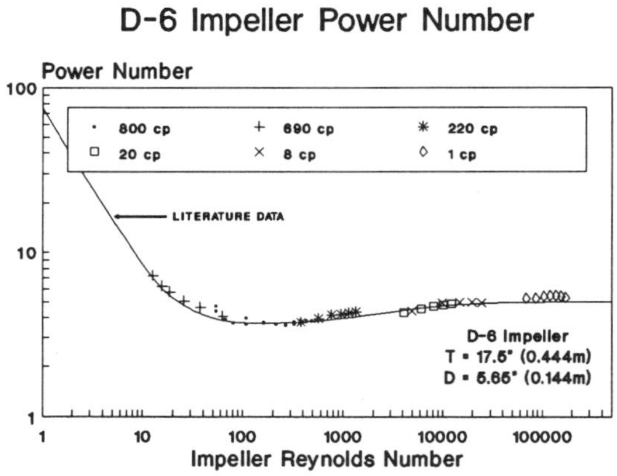

Figure 9

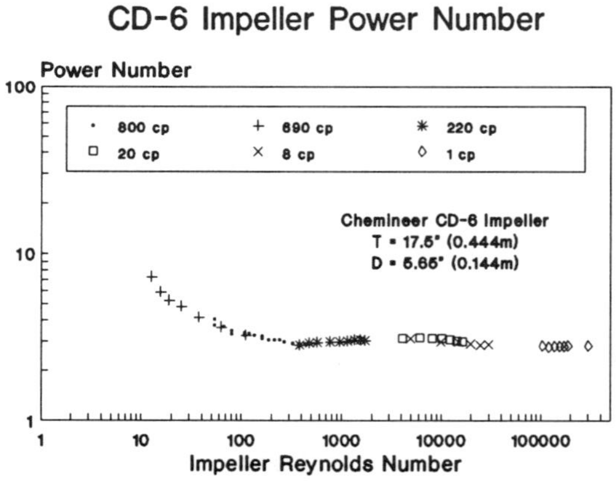

Figure 10

Gassed Power Draw of a D-6 Impeller

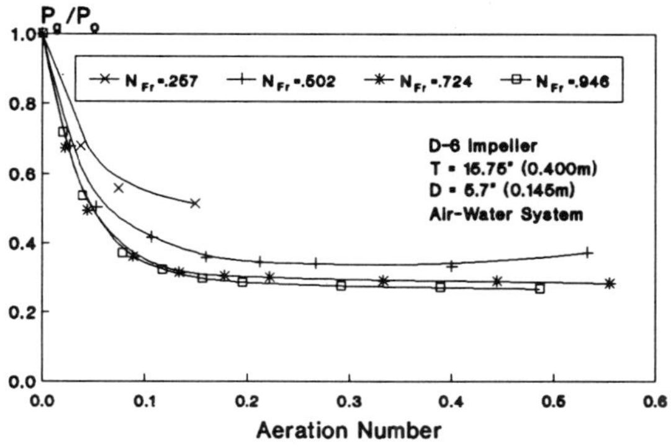

Figure 11

Gassed Power of a CD-6 Impeller

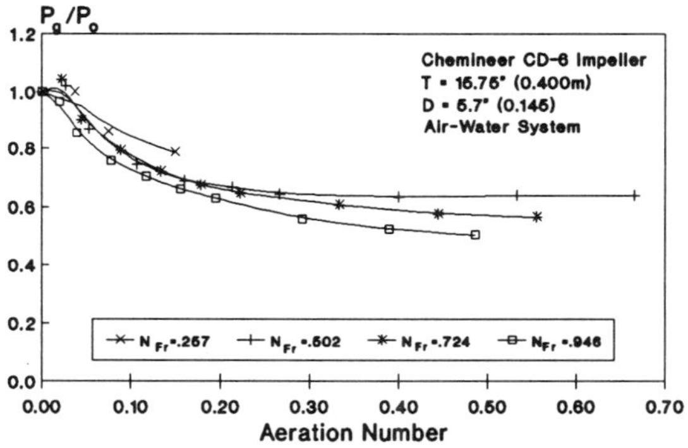

Figure 12

Gassed Power Comparison Between D-6 and CD-6 Impellers

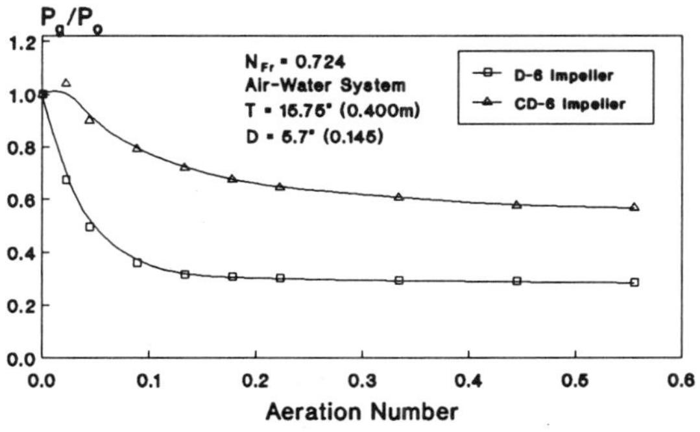

Figure 13

Viscous Effects on Gassed Loading of D-6 Impeller

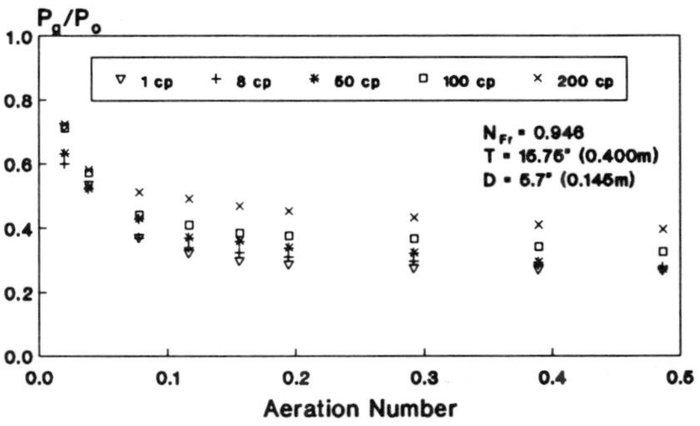

Figure 14

Viscous Effects on Gassed Loading of CD-6 Impeller

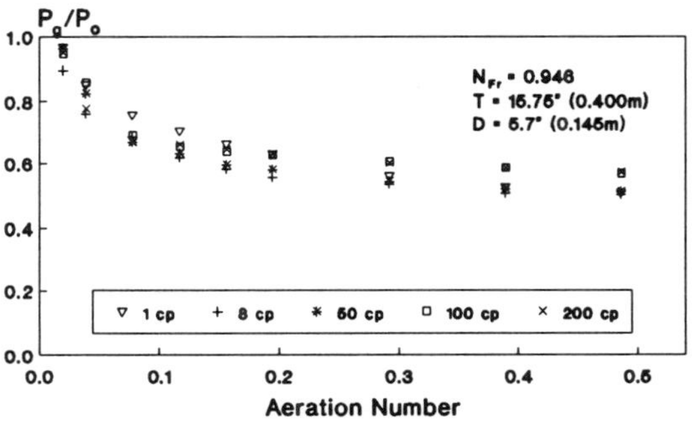

Figure 15

Holdup for a D-6 Impeller in an Air-Water System

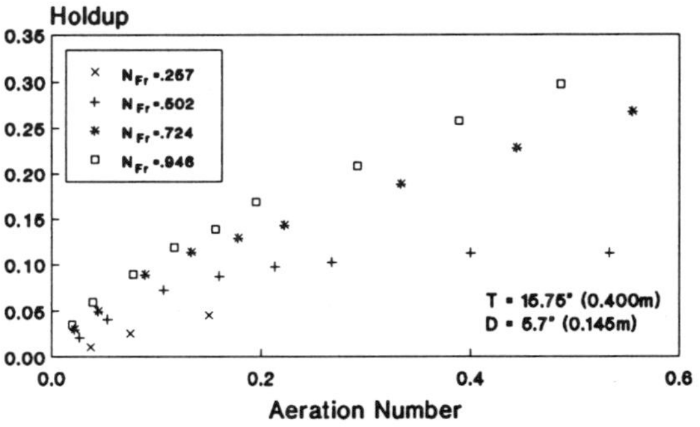

Figure 16

Holdup for a CD-6 Impeller in an Air-Water System

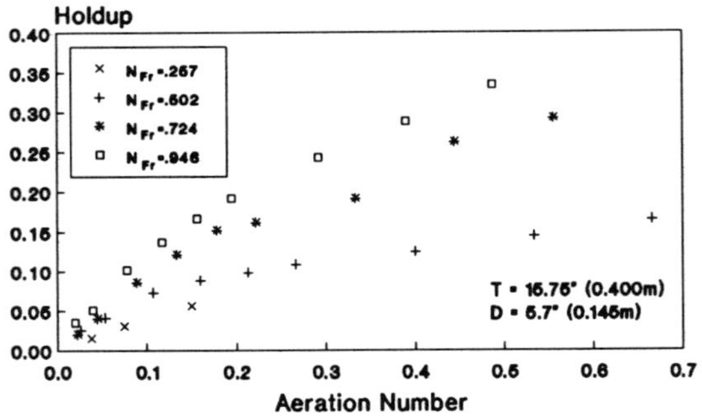

Figure 17

Holdup for a CD-6 Impeller at Various Viscosities

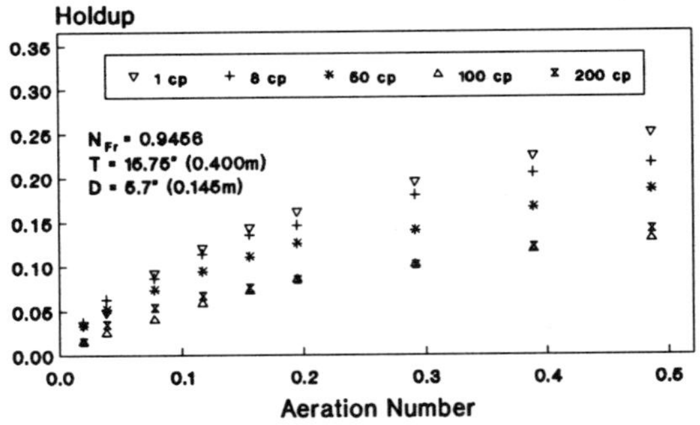

Figure 20

Holdup Comparison of D-6 and CD-6 Impellers

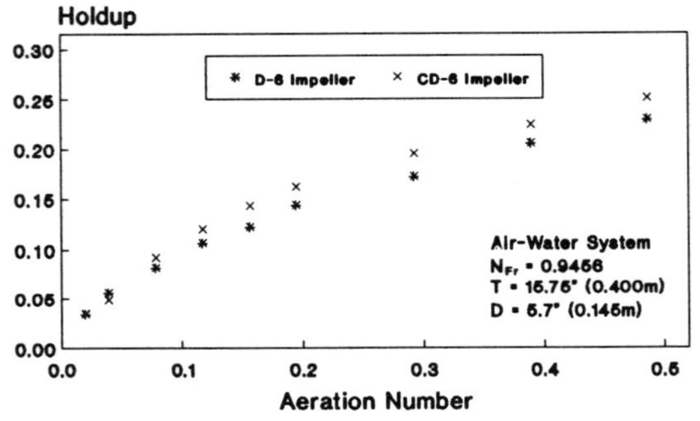

Figure 18

Plot of Holdup Data Using Smith's Correlation

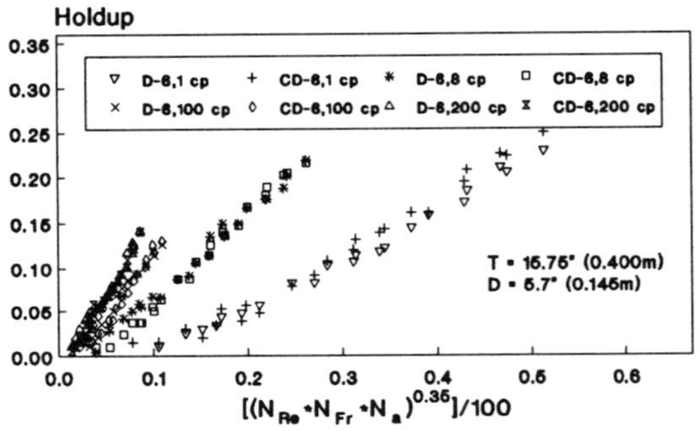

Figure 21

Holdup for the D-6 Impeller at Various Viscosities

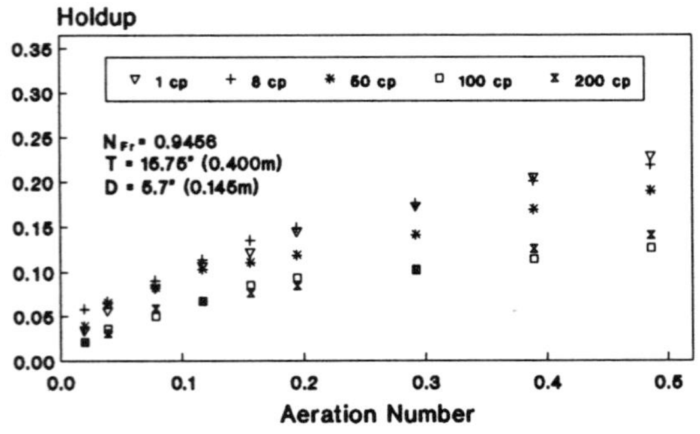

Figure 19

Holdup Using New Correlating Parameter

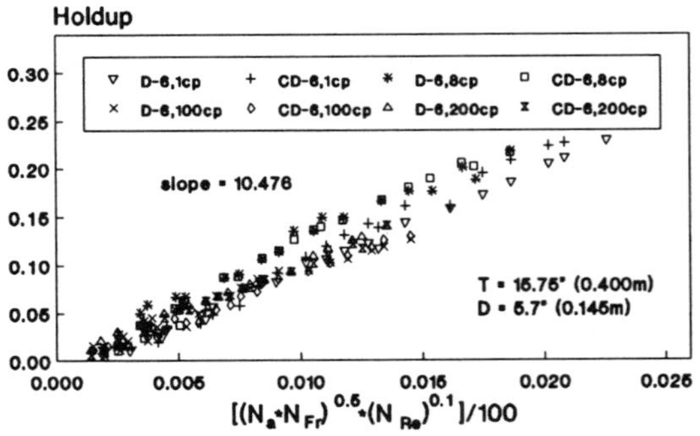

Figure 22

Flooding Characteristics of Impellers in Stirred Tanks

R. N. Sharma, R. A. Howk and K. S. Lally

Lightnin, Rochester, NY 14611

A fresh look and a critical examination of the flooding characteristics of radial, axial and mixed flow impellers for an air-water system in stirred tanks forms the basis of the present study. The ratio of gassed to ungassed power was measured and plotted for increasing speeds at constant gas rates in a 0.44 m baffled tank with D/T of approximately 1/3. The curves reveal two minimas and two maximas. The whole dispersion field, therefore, can be split at four recognizable points, the complete flood point N_{CF}, the partial flood point N_{PF}, the dispersion point N_D and the dispersion with recirculation point N_{DR}. The determination of power profiles at constant gas rates provides a non-invasive measuring technique and allows for rigid definitions for the four dispersion points, which hitherto had not been well established. The results suggest the causes of existing discrepancies in the reported correlations which need to be modified to bear universal validity. Suggested nomenclature for the points of interest is N_F (for N_{CF}) and N_R (for N_{DR}), and N_D.

INTRODUCTION

In gas-sparged stirred tanks, the gas tries to impose its flow pattern on the liquid, while at the same time the rotating mixer also tries to generate its flow pattern in the stirred liquid charge. Two extremes can easily be distinguished in this context: 1) At low impeller speeds, the mixer is rendered ineffective by losing its pumping capacity and the buoyancy induced gas-flow controls the liquid circulation. The impeller at this condition is said to be flooded. 2) A role-reversal takes place at higher speeds where the mixer controls the liquid flow pattern with its pumping capacity progressively restored depending on the rate of gas flow. It becomes, therefore, imperative to identify these two extremes which correspond to flooded and dispersed conditions of the sparged gas.

It can be intuitively argued that the successive stages between the flooded and dispersed conditions appear to be continuous. The identification techniques for the two conditions are often subjective (visual observations), or not truly responsive to the local flow character (physical probes). This gives rise to variance in the experimental data and, therefore, in the resulting correlations. The situation is compounded further when the definition of the flooded state of the impeller is loosely defined. The purpose of the present study is to establish, without any ambiguity, various recognizable points (including the flood point) in the entire dispersion flow-field. A critical examination of the dispersion phenomena is probably required to set universal definitions of different dispersion points and to focus attention on the corresponding correlations. It is not the intent of the present work to evaluate the existing correlations or to recommend one, except to suggest that the discrepancies in the correlations could be due to the lack of firm and rigid definitions of various dispersion points and the measuring techniques to identify them.

In most of the earlier investigations, radial flow impellers have been used to disperse the gas in stirred tanks. Initially, attention was devoted to the flood point, which, unfortunately, was defined in two ways. One, in line with the observations of Rushton and Bimbinet (1968), defined flooding as the state where the bubbles ceased to travel radially out along the impeller discharge stream. This can be identified as a transition from b to c in Figure 1 which is a typical characterization of the bulk dispersion, as given by Nienow et. al. (1977), both through photographic observations and through a P_g/P_0 versus Q/ND^3 curve. The second definition is due to Nienow et. al. (1977) who defined the onset of flooding as the point where the bubbles just started to circulate along the lower recirculation loop of the liquid discharge stream. This state, which can be identified as a transition between Figs. 1(c) and 1(d), evidently encompasses a larger volumetric space than the flood-point dispersion of Rushton and Bimbinet. However, both of the definitions were denoted as N_F. Later, to remove any ambiguity, at least in nomenclature, Nienow's flood point has been renamed as N_{CD}. Whether by design or coincidence, this definition seems to have an added advantage in that N_{CD} corresponds to the minimum in P_g/P_0 versus N curves. In fact, in the present study, such curves were generated for different dispersed systems in order to identify various dispersion points on the curves for radial, mixed flow and axial flow impellers. Visual observations and photographic evidence were also used to supplement the identification process.

EXPERIMENTAL

Figure 2 is a sketch of the test assembly. Experiments were conducted in an acrylic tank of diameter 0.44 m., fitted with standard baffles. Air, after passing through a rotameter, was sparged through a ring sparger placed concentrically below a centrally mounted mixer. A standard Rushton turbine, 45^0 4-Pitched Blade Turbine and a 1.5 pitch Marine Propeller were used. The D/T ratio was kept constant at 0.34. For a specified gassing rate, the impeller speed was progressively increased and for each speed, power was measured. Signals from speed and torque sensors were averaged and recorded by a PC through a data acquisition assembly. Over 500 points were collected in a sample and the averaging was extended to 120 such samples. In fact, the speed and torque signals were "over averaged". This was done to ensure repeatability of the data which was within 5 percent. Various states of dispersion were visually observed and photographed.

RESULTS & DISCUSSIONS

Standard Rushton Turbine. A typical power profile for a Rushton Turbine at constant gas flow rate is shown in Figure 3. Let us first analyze and identify various recognizable dispersion points on this curve for a typical gas flow rate of 0.0009 m^3/sec. The interpretations will later be extended to higher flow rates to check if these points retain universality in their occurrence. In fact, this process can be repeated in the future to cover a wide range of D/T also. It is more logical to extend the study to other D/T's once an appropriate understanding has been developed for the constant D/T case. Four points can easily be recognized in Figure 3(a) or in Figure 3(b) where the x-axis has been changed from N to Q/ND^3 (N increases towards left). Referring to Figure 3(b), the impeller remains completely flooded until the speed increases to N_{CF}, the first minima. At this point the impeller has entered the pumping mode and just starts to deflect the vertical geysers of gas. The gassed power, up to this point, decreases more rapidly than the ungassed power with the result that the ratio P_g/P_0 decreases with increasing N at a constant gas rate. This corresponds to stage ii of bulk dispersion. Note that the usual bubble flow patterns as given by Nienow et. al. (1977) in Figure 1 are slightly modified here. The stage where the upper secondary recirculation loop occurs has been eliminated. Instead, one more stage has been added, in their representation, between (a) and (b) which is stage ii, the complete flood point.

Beyond N_{CF}, as the impeller pumping increases, the gassed power increases with N more than the ungassed power up to N_{PF}. N_{PF} can be recognized as the partially flooded state. This corresponds to stage (iii). P_g/P_0 rises as N is increased from N_{CF} to N_{PF}. Also, the impeller, during this increase, builds up its pumping capacity. At N_{PF}, the impeller enters the dispersion mode. A discernable curvature or profile is set-up along the outer contours of the dispersed region. Further increases in impeller speed progressively increase this curvature until point N_D, where the curvature becomes infinite and complete dispersion above the impeller is achieved. Stage (iv) corresponds to this state of dispersion. It should be noted that the contours of this increasing curvature may not be distinctly identifiable since the smaller bubbles from a wide bubble spectrum blur it. This is the most important point in the dispersion field. Even up to date, it has often been erroneously identified as the flood point. It is intuitively suggested that it is always identifiable through the power response curve. In the present case of medium D/T, it corresponds to the second minima.

The question as to why the ratio P_g/P_0 between N_{CF} to N_D goes through a maxima at N_{PF}, can best be answered in terms of the changes in cavity structure during this range of speed. It is, however, not the present objective to define the shape of the curve in terms of cavity formation, its growth and structure. It would suffice to add that the impeller characteristics change from N_{CF} onwards. In fact, from N_{PF} to N_D, the changes in the impeller characteristics due to cavity structure are such that the impeller becomes more efficient because of progressively reduced drag. With still further increases in impeller speed, P_g/P_0 rises again until the point N_{DR} where dispersion with recirculation starts affecting the gassed power. The impeller becomes comparatively less efficient due to a change in cavity structure which increases the drag now.

N_D is the point which Nienow et. al. (1977) has named as N_{CD}. At this point, the gas is completely dispersed in the stirred region above the impeller. Nienow et. al's representation of power response curves (Figure 1) are similar to Figure 3(b) except that the portion N_{CF} - N_{PF} - N_D is missing. Instead, it is represented by a smooth curve directly connecting N_{CF} - N_D. It is possible that the former part of the curve might have been lost in the scatter of data if the torque measurements were not very sensitive in this low torque region. Thus, for the given two-phase stirred system, various dispersion points can be distinctly identified through the measured power response curves. Through visual observations and photographs, it was confirmed that N_D does represent complete dispersion in the stirred tank space above the impeller.

It is interesting to note that two additional points proposed earlier, besides N_{CD}, can also be identified in the power response curve of Figure 3. One, by Nienow et. al. (1977) as N_R (Figure 1), is the point above which the sparged gas recirculates back in the feed stream. This is expressed as N_{DR} in the present work and corresponds to the second peak in Figure 3. In fact, by placing a mirror below the tank, it was visually observed that recirculation starts earlier and progressively increases. At N_R, it seems to affect the gassed power. The power response curve offers a range of speeds beyond the second maxima where the recirculation point can be located. Tentatively, the maxima itself is taken as the point of recirculation. The point to be made is that, like N_D, the recirculation point can also be distinctly identified through the power response curve under discussion. The state of gas dispersion at N_{DR} is shown as 3(b)(v), but does not seem to be specific like other dispersion states.

The second point N_F, was defined by Nienow et. al. (1985) as the flood point, a transition from (a) to (b) in Figure 1. It corresponds to a step increase at N_F in the plot of P_g/P_0 versus Q/ND^3 at a constant impeller speed rather than a constant flow rate. It is, therefore, unclear whether N_F corresponds to N_{CF} or N_{PF}. From the sketches of the bulk flow dispersions, it matches the descriptions of N_{CF}.

Thus far, we have traced the existence of four dispersion points on the P_g/P_0 versus N (or Q/ND^3) curve at a given gas rate and geometry. To study the effect of diverse flow rates on the existence of the four dispersion points, P_g/P_0 versus N curves were generated at various flow rates as shown in Figure 4. In general the curves display two minimas and two maximas as discussed earlier.

To add credence, experimental verification of the present observations is sought from literature. Ismail et. al. (1984) did report similar findings but somehow their observations got obscured in the literature. Figure 5 represents their power response curves at constant flow rates. They also observed two minimas and two maximas and recognized the same dispersion points as in the present case along dotted curves AA', BB', CC', and DD'. A disc turbine with eight blades was used. In fact, the authors observed that a rigid definition of flood point would be along the locus of the first minima (N_{CF} of the present work) although they added that a convenient definition of flood point would be along the locus of the second maxima defined as N_D in the present work. It seems more logical to stick to rigid definitions since approximate or convenient definitions may lead to erroneous results as the observations based on these definitions may not follow a predictive course when the experimental conditions are widely changed. Recently, Svihla et. al. (1992) conducted similar experiments on two close tank sizes, 0.241 m and 0.298 m. Their findings too, are similar to the present observations. Figure 6 shows their P_g/P_0 versus N curves at different flow rates for the 0.298 m tank. Four dispersion points corresponding to two minimas and two maximas can easily be recognized.

It can be observed from the present set of curves and the ones from the two cited references that N_{PF} and N_{DR} (i.e., the two maximas) are not clearly identifiable at higher gas flow rates. They seem more to be camouflaged rather than becoming extinct. It may be possible to develop an explanation of this behavior in terms of cavity structure and its growth. More experiments are in progress on this aspect. In the meantime, extrapolated values of N_{PF} and N_{DR} will be used at high flow rates.

Comparisons for the different critical speeds of gas dispersion from the present study and a group of reported investigations for T = 0.44 m and D = 0.15 m are shown in Figure 7. Likewise, Figure 8 shows the same type of comparisons with the data of Svihla et. al. (1992). Experimental data of Ismail et. al.(1984) and Weinspach et. al. (1981) are also included in Figure 7 since their experimental conditions are similar to the present study. It is interesting to note that the data of Weinspach et. al. (1981) almost exactly follow the corresponding data of N_D from the present experiments. The N_D data of Ismail et. al. (1984) expectedly falls below the present study since they used an 8-blade turbine. Curves 3 to 9 seem to encompass the present data for N_{CF} and N_D. It would have been logical to analyze the reasons for mismatches if the dispersion points were identified through the power profiles as was done in the present experiments. Additionally and more importantly, the definition of various dispersion points should have been consistent. Unfortunately, as pointed out earlier, rigid definitions and faithful measuring techniques had not yet evolved. Most of the attention has been focused on correlating N_D (for example lines 3, 4, 5, 8 and 9). The associated measuring techniques employed were visual observations, both simple and derotational, power response curves, micro-propeller response and hot wire anemometry. Superimposed on these uncertainties is the fact that most of the earlier workers missed N_{CF} and N_{PF} points. These points need to be defined and identified lest an erroneous mixup may result if an N_D - correlation is directly attempted.

N_{CF}, representing a flooding-loading transition, line 8, was originally defined by Smith & Warmoeskerken (1985) as the cessation of radial pumping of the impeller. It matches the Rushton-Bimbinet definition of flood point. Later, Warmoeskerken and Smith (1985) measured the flooding-loading transition by using a micro-propeller similar to Rouston and Bruxelamane (1981). Evidently, a dispersion point below N_D was measured, probably between N_{CF} and N_{PF}. Nienow et. al. (1985) seem to confirm this point. They revised (line 7) the Smith and Warmoeskerken (1985) correlation. It is interesting to note that line 7 and 8 closely follow the N_{CF} and N_{PF} data in Figures 7 and 8. Lines 1 and 2 on these figures refer to N_{DR}. The present data and that of Svihla et. al. (1992) fall in between these two lines. The basic reason for this satisfactory agreement is that N_{DR} has been identified through power response curves in all these studies.

A detailed analysis of the variance in the correlations, between themselves or in comparison with the present data, would not be, as pointed out earlier, very meaningful if a consistent definition is not followed. Figures 7 and 8, therefore, highlight the variance only in this context. For radial flow impellers, the present study suggests rigid definitions of various dispersion points which can be identified through non-invasive measurements of power profiles. It is recommended that the study of N_{PF} be tentatively dropped to focus attention on the remaining three points N_{CF}, N_D and N_{DR} along which the dispersion field appears to be split. If N_{PF} is not considered significant, the nomenclature of the three points could be revised to the flood point N_F, the dispersion point N_D and the recirculation point N_R. Results of the work in progress will be reported to confirm if these points can still be identified at diverse values of D/T ratio.

Mixed Flow and Axial Flow Impellers. Preliminary work was conducted to ascertain if the power response curves could also be used to identify various dispersion states for mixed and axial flow impellers. Power profiles for two gas flow rates, one low and one high, were generated for PBT-4 and Marine Propeller at a D/T of approximately 0.3. These are shown in Figures 9, 10, 11 and 12. The curves closely resemble those reported in the literature, for example, due to Rewatkar and Joshi (1991) for PBT-4 and Chapman et. al. (1983) for both PBT-4 and Marine Propeller. At the left part of the curves, the rate of decrease of the gassed power with increasing speed is more than the rate of the ungassed power with the result that P_g/P_0 decreases to a minima. This part of the behavior is similar to the Rushton Turbine up to the first minima. Both PBT-4 and Marine Propeller at this point enter into the pumping mode and try to reverse the direction of the upward flush of the gas by imposing flow patterns compatible with their downward pumping action. A kind of instability is set-up at this point when the state of dispersion oscillates between completely dispersed and flooded states with a minimum periodicity in terms of the occurrence of the completely dispersed state. This minima can be regarded as the flood point. The oscillations continue until the maxima is reached after a relatively small increase in speed. The periodicity of the completely dispersed state is maximum here. The point representing the maxima of the short unstable region assumes importance as the point of complete dispersion denoted as N_D. Correlations can be developed for this point.

At low gas flow rates, the dispersion point at the maxima can easily be located in Figures 9 and 11. At higher flow rates, however, the maxima is not as sharp and distinct. In fact, at very high flow rates, the curve becomes flatter beyond the flood point and the maxima, being camouflaged, can not be easily identified. To obviate the difficulty of locating N_D, particularly at high flow rates, one can utilize the response of the torque fluctuations with speed. Figures 9, 10, 11 and 12 also show the plot of the torque variance, defined as the ratio of the difference between the maximum and minimum torque to the average torque for any speed at a constant gas rate. Expectedly, the value of this variance is maximum at the complete dispersion point. Interestingly, at higher flow rates, while the power profile becomes flatter, the torque curve becomes more sensitive and sharp. This offers an alternate method of accurately measuring N_D as is evident from Figures 10 and 12.

Figure 13 represents the curves of torque variance for all the impellers. The corresponding power response curves are shown in Figure 14. Evidently, torque variance curves provide yet another non-invasive technique to identify various dispersion points for all the impellers, including the Rushton Turbine. Chapman et. al. (1983) defined the speed for complete dispersion ahead of the maxima where the torque fluctuations are more steady. This is to avoid an undesirable state of large torque fluctuations. The apparent increase in N_D may be regarded as a form of safety factor which needs to be incorporated in to the final correlation and not during its development. The decrease in the torque variance is not sudden, as is obvious from Figures 9 - 12, and the point at which the torque variations are steady enough remains subjective. Measurements of the torque variance curve in different systems may provide a clue for the safety factor.

CONCLUSIONS

The dispersion field in a sparged stirred tank with a Rushton Turbine has been shown to split at four recognizable points when viewed on P_g/P_0 - N (or Q/ND^3) curves at constant gas rates. This provides rigid definitions and a measuring technique for the four dispersion points, N_{CF}, N_{PF}, N_D and N_{DR}. However, attention on N_{PF} representing the partially flooded state can tentatively be reserved. The remaining three points can be renamed as N_F the flood point, N_D the dispersion point (above the impeller) and N_R as the recirculation point.

For the mixed flow and axial flow impellers, the complete dispersion point N_D corresponds to the maxima in the P_g/P_0 versus N curves. A curve of torque fluctuations with N provides an alternate non-invasive technique to locate N_D, particularly at high flow rates. This can also be applied to identify the dispersion points for Rushton turbine. The results need to be extended to larger scales with diverse D/T ratios to examine the response of various dispersed conditions discussed above.

NOTATION

D	=	impeller diameter
N	=	impeller speed
N_D	=	dispersion point
N_{CF}	=	complete flood point
N_{PF}	=	partial flood point
P_g	=	gassed power
P_0	=	ungassed power
N_{DR}	=	dispersion with recirculation point
N_R	=	recirculation point
N_F	=	flood point
T	=	tank diameter
T_{max}	=	maximum torque
T_{min}	=	minimum torque
T_{av}	=	average torque

LITERATURE CITED

Chapman, C.M., A.W. Nienow, M. Cook, and J.C. Middleton, "Particle-Gas-Liquid Mixing in Stirred Vessels Part II: Gas-Liquid Mixing, Chem. Res. Des., Vol.61, March 1983.

Greaves, M., and K.A.H. Kobbacy,"Power Consumption and Impeller Dispersion Efficiency in Gas-Liquid Mixing", Fluid Mixing, I. Chemical Engineering Symposium Series, No. 64, I. Chemical Engineering, Rugby, Warks, England, L1, 1981.

Ismail, A.F., Y. Nagase and J. Imon, "Power Characteristics and Cavity Formation in Aerated Agitations", AIChEJ, 30, 487, 1984.

Lu, Wei-Ming, and Hsuan Chen, "Flooding and Critical Impeller Speed for Gas Dispersion in Aerated Turbine-Agitated Vessels", Dept. of Chem. Eng., National Taiwan Univ., Taipei, Taiwan, -- (unknown).

Nienow, A.W., M.M.C.G. Warmoeskerken, J.M. Smith, and M. Konno, "On the Flooding/Loading Transition and the Complete Dispersal Condition in Aerated Vessels Agitated by a Rushton Turbine", Proc. 5th European Conference on Mixing, Wurzburg, Germany, BHRA Fluid Engineering, Cranfield, England, 143, 1985.

Nienow, A.W., D.J. Wisdom, and J.C. Middleton, "The Effect of Scale and Geometry on Flooding, Recirculation, and Power in Gassed Stirred Vessels", Proc. 2nd European Conference on Mixing, Cambridge, England, BHRA Fluid Engineering, Cranfield, England, F1-16, 1977.

Rouston, M., and M.Z.A. Bruxelmane, "Loading and Flooding Points in Turbine Agitated Gas-Liquid Dispersions", 7th International Congress of Chemical Engineers, CHISA, Prague, Czechoslovakia, B3.3, 1-8, 1981.

Rewatkar, V.B., and J.B. Joshi, "Role of Sparger Design in Mechanically Agitated Gas-Liquid Reactors Part I: Power Consuption", Chem. Eng. Technol., 14, 386, 1991.

Rushton, J.H., and J.J. Bimbinet, "Holdup and Flooding in Air Liquid Mixing, Can. J. Chemical Engineering, 46, 16 (1968).

Svihla, C. K., and co-workers, Chem. Eng. Dept., Univ. of Louisville Speed Scientific School, Private Communication, 1992.

Smith, J.M., and M.M.C.G. Warmoeskerken,"The Dispersion of Gases in Liquids with Turbines", 5th Euro. Conf.on Mixing, Wurzburg, West Germany, 13, 1985.

Warmoeskerken, M.M.C.G., and J.M. Smith, "Flooding of Disc Turbines in Gas-Liquid Dispersions: A New Description of the Phenomenon", Chem. Eng Sci., 40, 2063, 1985.

Wiedmann, J.A., A. Steiff, and P.M. Weinspach, "Experimental Investigations of the Fluid Dynamics in Stirred Aerated Slurry Reactors", 7th. Int. Cong. Chem. Eng., Chem. Equip. Des. and Autom., Czechoslovakia, Aug.31-Sept 4, B3.2, 1981.

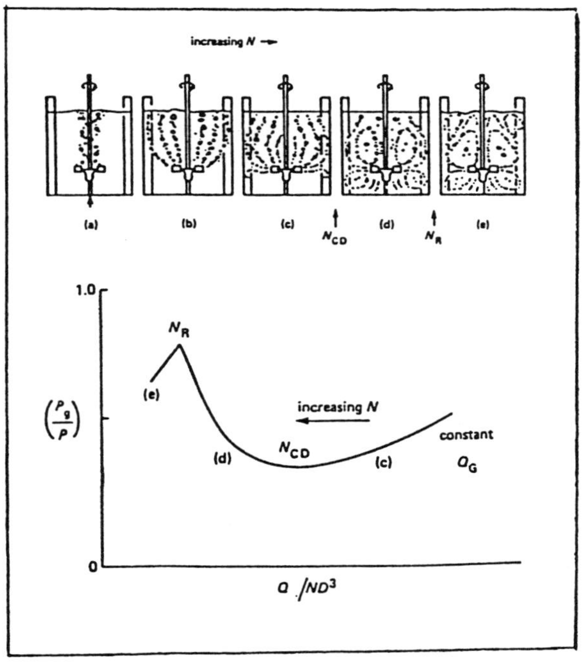

Figure 1 Power response curve and bulk flow patterns at constant gas rate

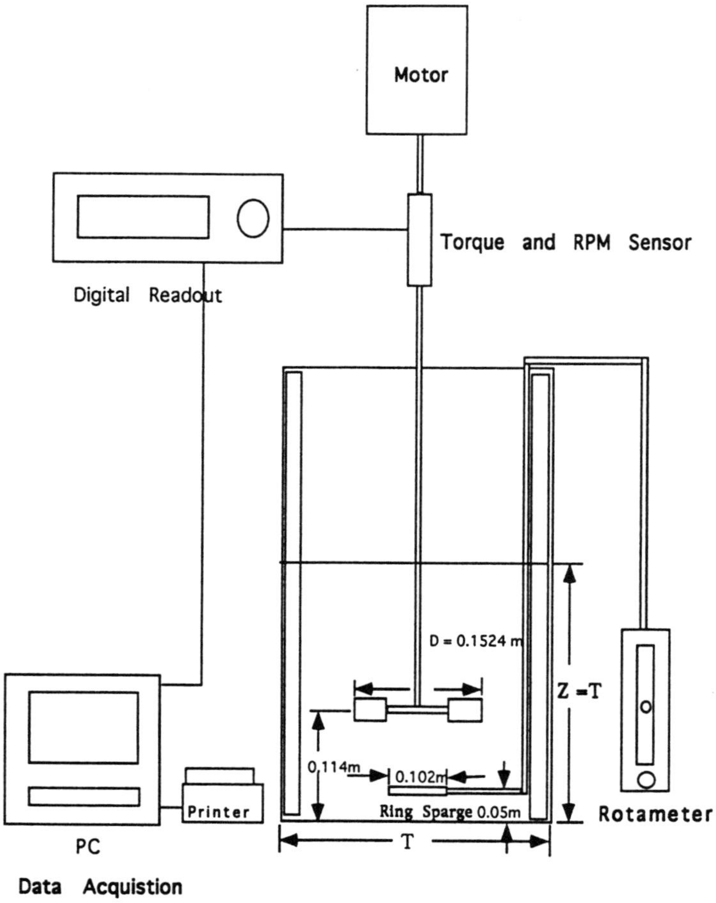

Figure 2 Experimental set-up

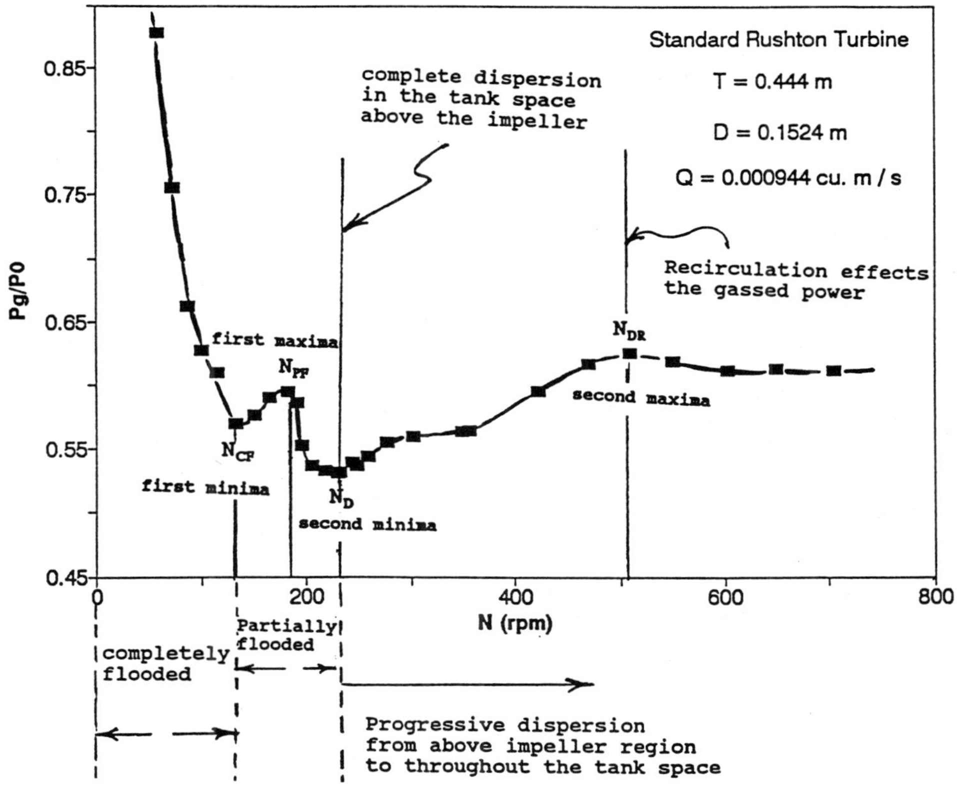

Figure 3(a)   Power versus speed at constant gas rate

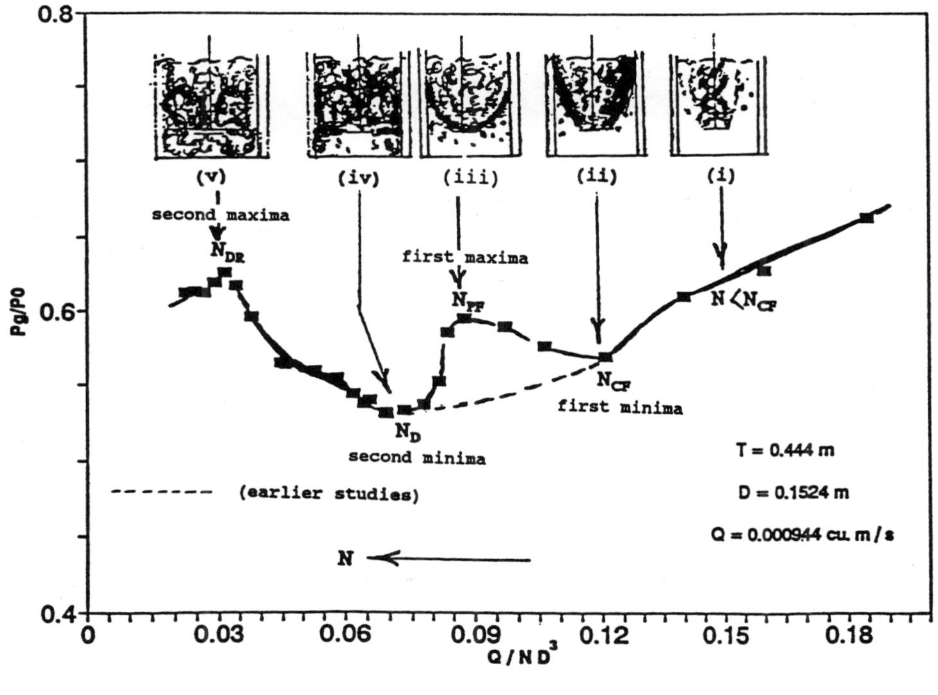

Figure 3(b)   Power versus aeration number at constant gas rate

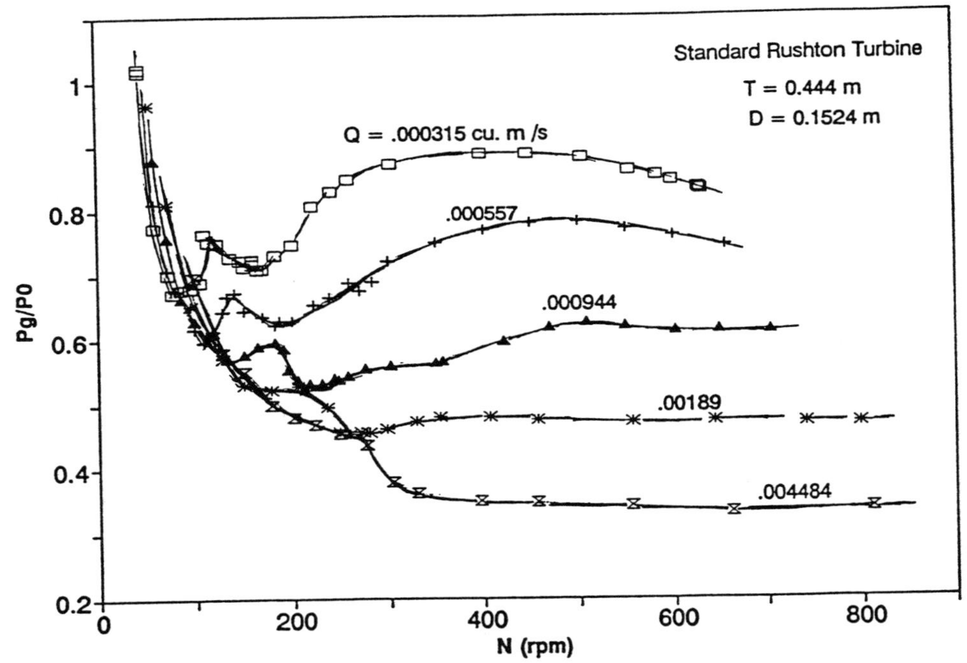

Figure 4 Power versus speed at constant gas rates

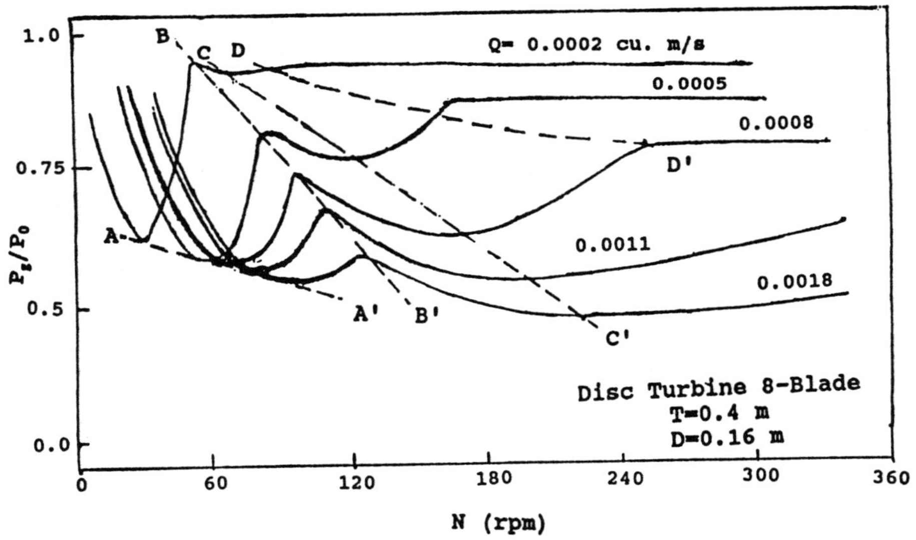

Figure 5 Power versus speed at constant rates (Ismail et.a.1984)

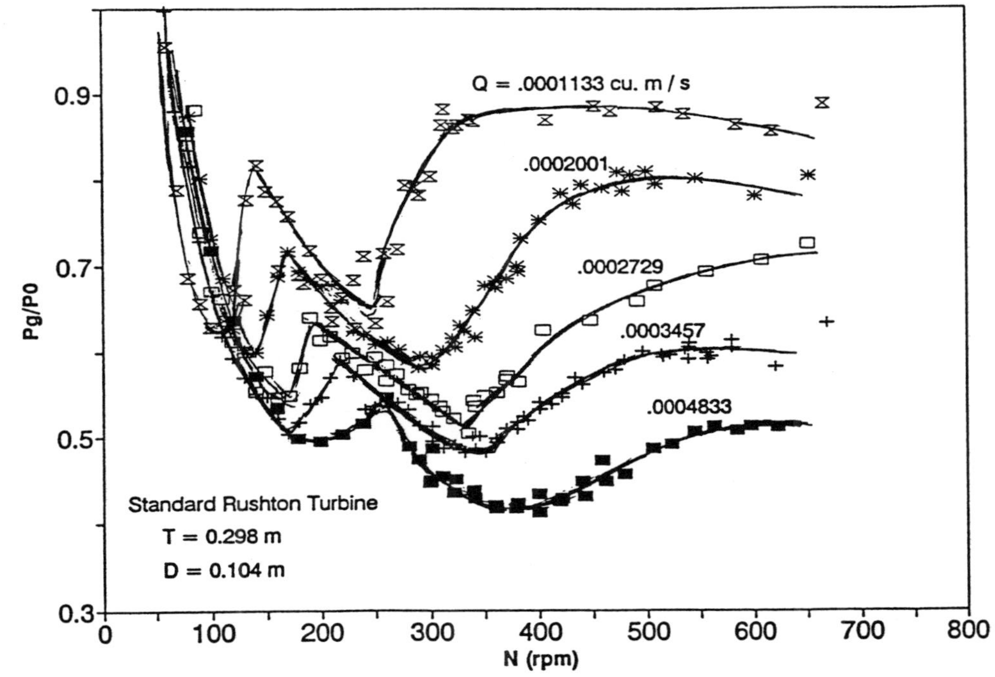

Figure 6 Power versus speed at constant gas rates (Svihla e.a.1992)

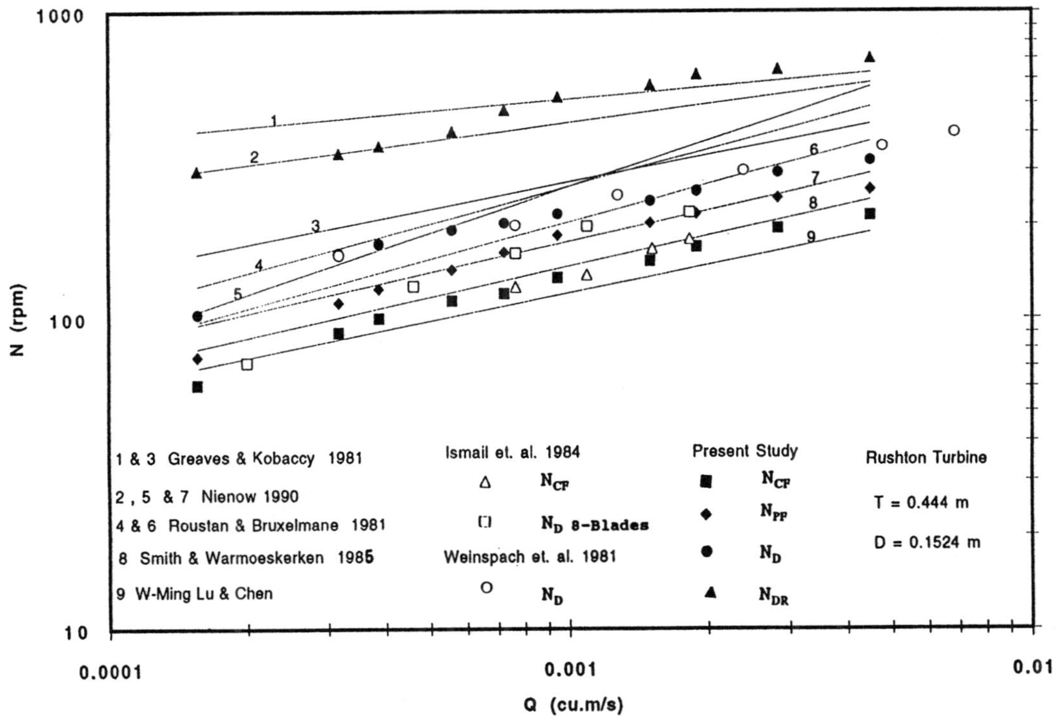

Figure 7 Comparisons between present data and correlations from literature

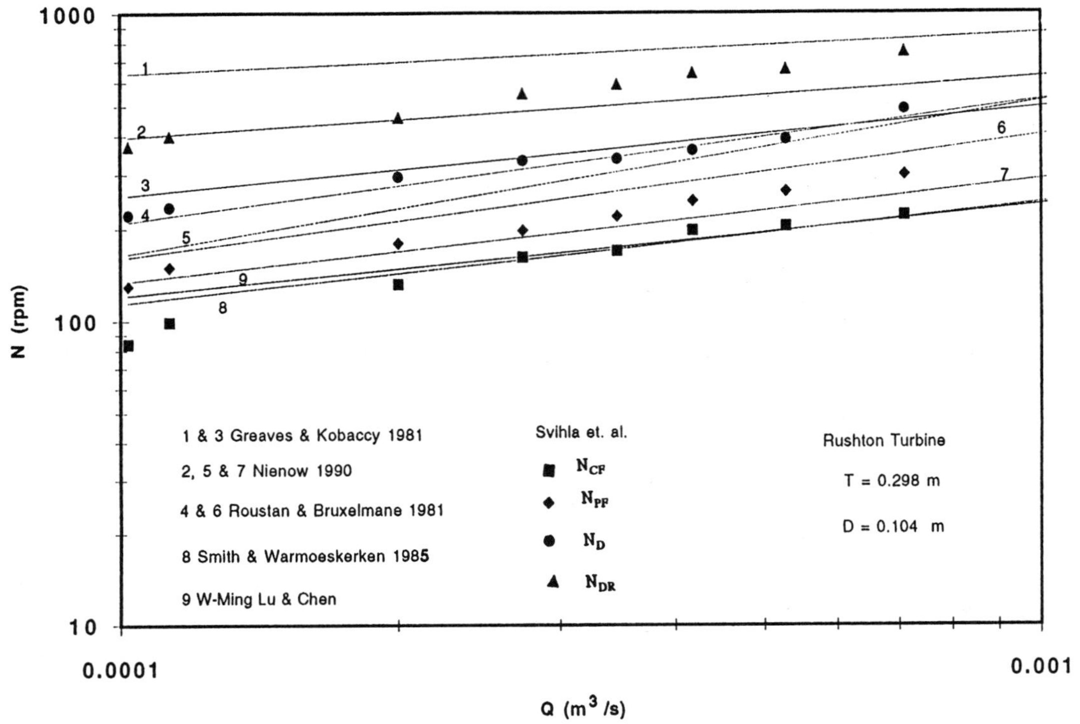

Figure 8 Comparisons between the data of Svihla et.al. (1992) and correlations from literature

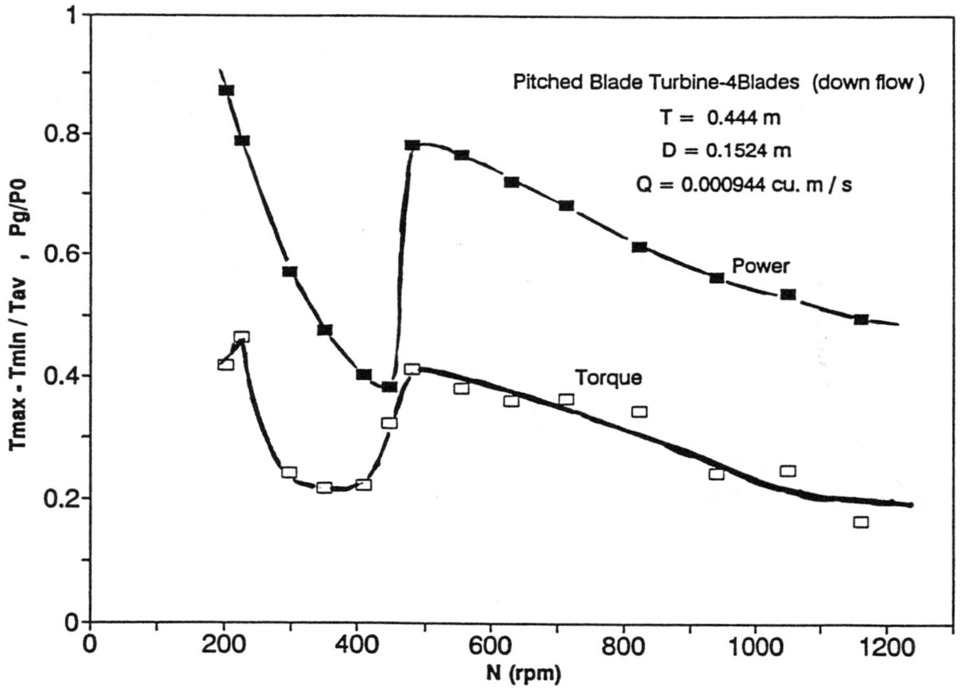

Figure 9 Torque fluctuations and power versus speed at a "low" gas rate for Pitched Blade Turbines

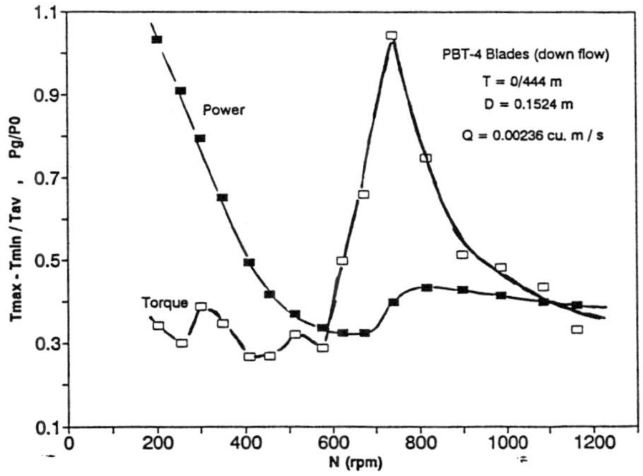

Figure 10 Torque fluctuations and power versus speed at a "high" gas rate for Pitched Blade Turbines

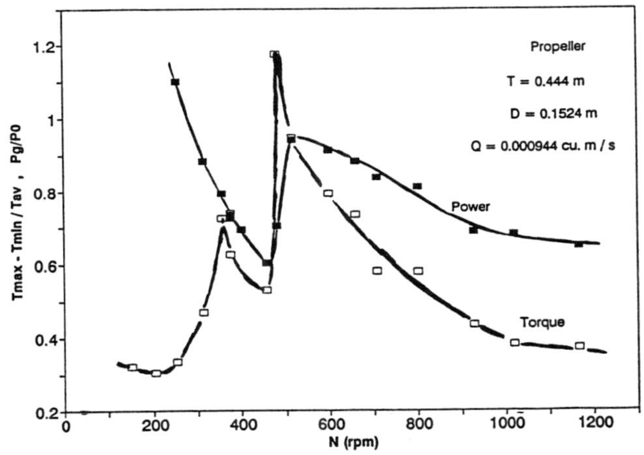

Figure 11 Torque fluctuations and power versus speed at a "low" gas rate for Marine Propeller

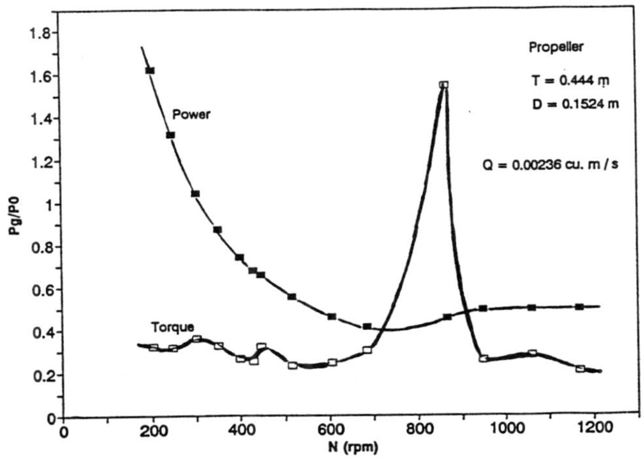

Figure 12 Torque fluctuations and power versus speed at a "high" gas rate for Marine Propeller

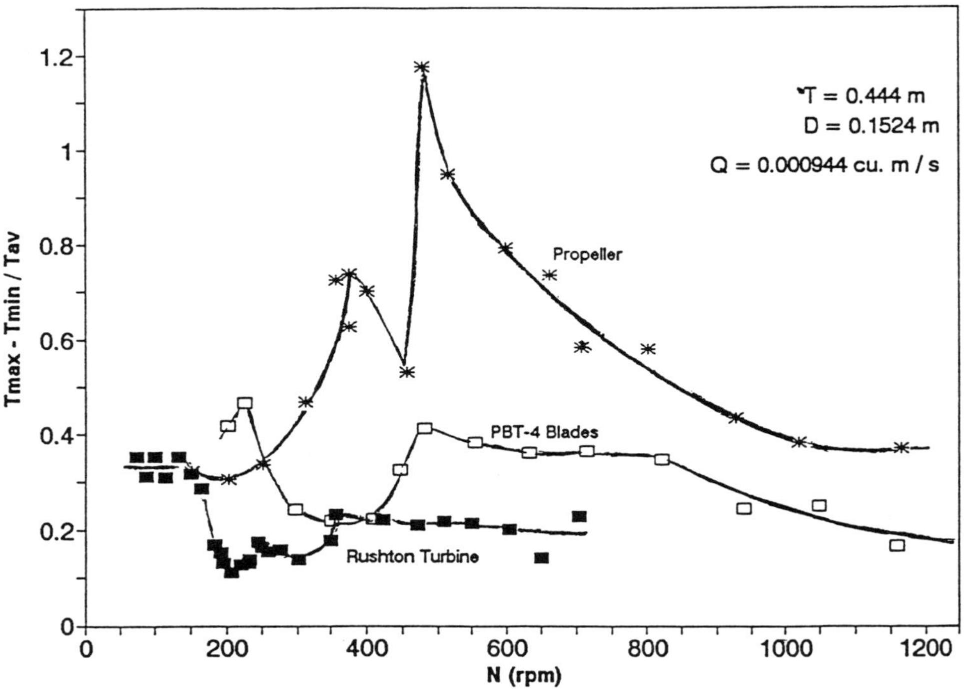

Figure 13 Torque fluctuations and power at a constant gas rate for all impellers tested

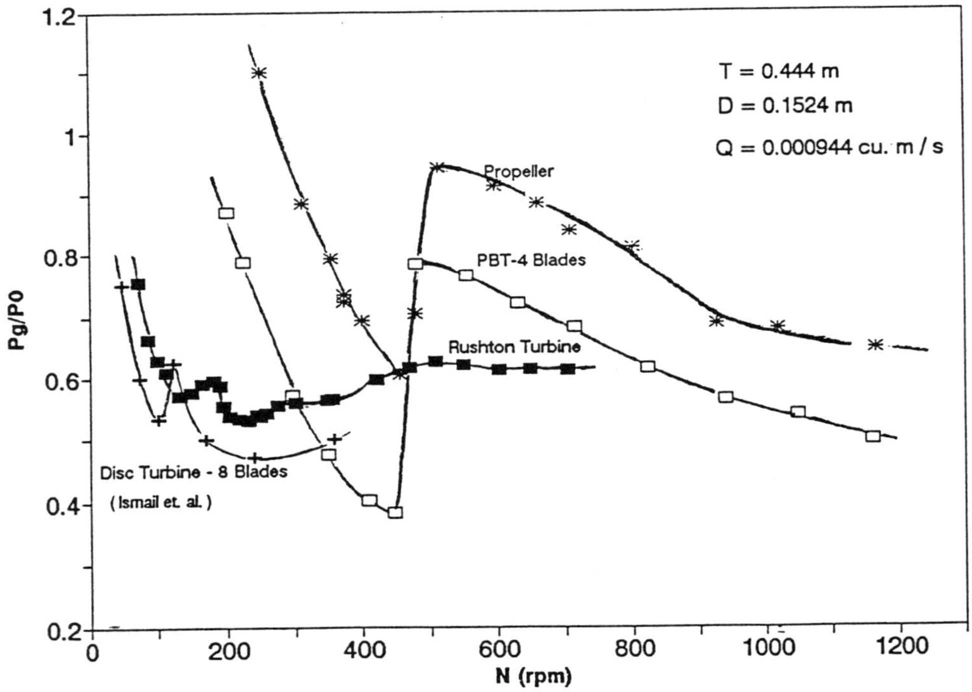

Figure 14 Power versus speed at constant gas rate for all impellers tested

Breakup Behavior of Bubbles in an Aerated Stirred Vessel

Rajarathinam Parthasarathy and Nafis Ahmed

Department of Chemical Engineering, The University of Newcastle, NSW 2308, Australia

Experiments are reported in which the bubble size and size distribution are measured as a function of agitation in an aerated mixing vessel. By maintaining non-coalescing conditions, it has been possible to monitor the breakup phenomenon in isolation. Bubbles of two different sizes, Sauter mean diameters of 300 μm and 2.5 mm are generated using porous spargers. The initial bubble populations are found to be well represented by log-normal distribution in both cases. With agitation, the 300 μm bubbles are too small to be broken further and the size distribution is preserved for the range of impeller speeds used. The larger bubbles are broken by agitation and the size distribution and the Sauter mean decrease. Changes in the bubble size distribution with step changes in the impeller speed are presented for the Rushton and 45 degree pitched blade turbines and marine propeller.

The most important factor in operations involving gas-liquid stirred vessels is gas dispersion by the impeller. In bio-reactors the effectiveness of gas dispersion determines other parameters like the gas holdup, interfacial area and mass transfer coefficient, and hence the efficiency of the overall operation. The gas dispersion characteristics of various impellers have been studied widely. Particularly, the two-phase flow regimes around a six-bladed Rushton turbine have been investigated in detail (van't Riet and Smith [1]; Bruijn et al.[2]). In general, the gas sparged into the mixing vessel is dispersed from trailing vortices of gas-filled cavities formed behind the impeller blades. Previous studies on gas dispersion have attempted to relate the two-phase flow patterns and the impeller power consumption to the formation of gas cavities of various types. However, the number of bubbles resulting from the dispersion process and their size distribution have received less attention. Bubbles are the basic constituents of the dispersion and their size distribution determines the gas phase residence time and the interfacial area for mass transfer. Thus, any theory attempting to predict the performance of gas-liquid stirred vessels would need to take into account the bubble size distribution in the dispersion.

There have been relatively few studies reported on the bubble size distributions in stirred vessels. Kawecki et al.[3] measured the bubble size distribution in the impeller region of a stirred vessel by withdrawing the dispersion outside the tank into a small transparent column of square cross section and photographing them. The average bubble size was found to decrease with an increase in impeller speed and increase with increase in gas flow rate. Barigou and Greaves [4,5] measured the bubble size distribution at various locations of a stirred vessel using a capillary suction tube. The bubble size distribution was found to vary with position and deviated significantly from a normal distribution. The bubble size distribution in the impeller region shifted towards the lower end of the bubble size scale with increase in impeller speed, indicating an increase of bubble breakup rate. In another recent study, Takahashi et al. [6] found that the bubble size distribution measured in the impeller region had a positive skewness towards the higher end of the bubble size scale. They also found that the bubble size distribution moved towards the lower size range with an increase in the impeller speed.

Although the above studies report that the bubbles in a stirred vessel break up with increase in impeller speed and their size distribution moves towards the lower end of the bubble size scale, a complete description of the breakup phenomena is not available. Stirred vessels in biochemical and chemical industries operate in a wide range of impeller speeds depending on the process and operating conditions. It is difficult to estimate the bubble size distribution for these conditions from the available information, and quantitative studies on the breakup behaviour of bubbles and variation in their size distribution as a function of agitation or power input are warranted.

Recently, Parthasarathy et al. [7] studied the bubble break up in an aerated stirred vessel for a range of impeller speeds using two different impellers and three spargers. However, the work was limited to studying the variations in the Sauter mean diameter of bubbles with increase in impeller speed. The present paper reports the nature of change in the size distribution of bubbles when subjected to a range of agitation. To observe the bubble breakup process in isolation, that is independently of coalescence effects, the present study has been conducted in a non-coalescing medium.

EXPERIMENTAL

A schematic of the experimental set-up used in this study is shown in Figure 1. Experiments were carried out in a standard 0.195 m diameter, flat bottomed, cylindrical tank. Three standard impellers were used; six-blade Rushton turbine (radial-flow, high power), six-blade downward pumping 45° pitched-blade disc impeller (mixed-flow, medium power), and a marine propeller (axial flow, low power).

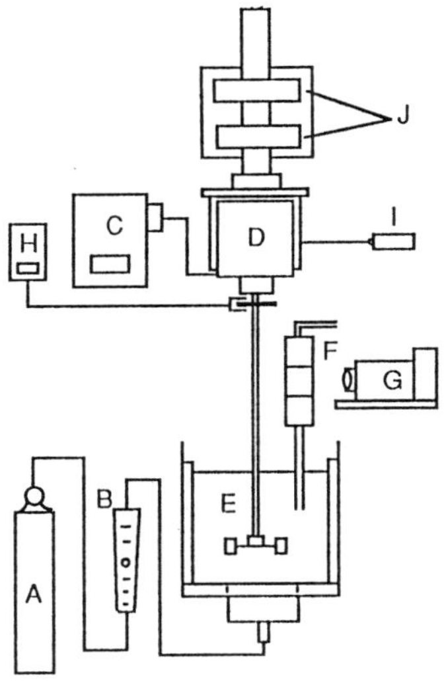

Figure 1. Schematic of the experimental arrangement. [**A**, Air cylinder; **B**, Rotameter; **C**, Variable speed drive; **D**, Motor; **E**, Tank; **F**, Photographic cell; **G**, Camera; **H**, Tachometer; **I**, Load cell; **J**, Bearings].

Bubbles of known sizes were generated by sparging air through inter-changeable, porous spargers placed at the tank bottom directly beneath the impeller. The mean size and the size distribution were then monitored as a function of the agitation in the vessel. To preserve clarity, results are presented for two bubbles sizes only, namely those generated using spargers incorporating pores of 150 to 250 μm and 5 to 15 μm in diameter. These spargers are designated as sparger-0 and sparger-4 respectively in the text, following the classification numbers used by the manufacturer (Pyrex). Water was used as the liquid phase and was made non-coalescing by adding 50 ppm v/v of methyl isobutyl carbinol, a common industrial frother. It was statistically confirmed that the bubble size distribution produced in the impeller zone was preserved in other regions of the tank. The surface tension of the liquid was measured to be 71.3 mN/m. The results presented are for a superficial gas velocity (U_g) of 1 mm/s. The top impeller speed for individual impellers was limited by the onset of considerable surface aeration, which was avoided for various reasons. The power drawn by the impeller was determined by measuring the torque experienced by the motor and impeller assembly unit using a load cell.

Bubble size was measured by photography. A stream of liquid containing a dispersion of bubbles was drawn into a specially-designed perspex photographic cell placed above the tank and photographed. This was made possible due to the non-coalescing nature of the medium. The photographic cell had two flat transparent windows between which the dispersion was drawn while being photographed. A 2-mm graticule pasted to the inside wall of the rear window was used as the calibrating scale. Sufficient number of bubbles were sized in each case to give statistically significant results.

Further details on the experimental arrangement is available in Parthasarathy et al. (7).

RESULTS AND DISCUSSION

Effect of Impeller Speed on Mean Bubble Size

In presenting the results, the bubble size distribution and the Sauter mean bubble diameter measured at 125 rpm are designated as the initial bubble size distribution and initial bubble size respectively. Below this speed, the bubbles rose vertically through the tank and the flow through the impeller region was influenced by the impeller type. The effect of impeller speed on Sauter mean diameter (d_{32}) for bubbles generated by the two relevant spargers are shown in Figure 2 for the Rushton turbine. The initial d_{32} of the bubbles generated by sparger-0 is about 2.5 mm, whereas the bubbles generated by sparger-4 are around 300 μm in diameter. It is observed that the 300 μm bubbles are unaffected by an increase in impeller speed and their size remains practically constant for the range of agitation used. The d_{32} for sparger-0 bubbles however, starts decreasing after 200 rpm and the value approaches that of sparger-4 bubbles at 800 rpm. Similar trends were observed for the other two impellers used.

Effect of Impeller Speed on Bubble Size Distribution

Sparger-0 bubbles. The size of the bubbles generated by sparger-0 and measured at the lowest impeller speed of 125 rpm varied from 1 to 4 mm. The frequency number distribution

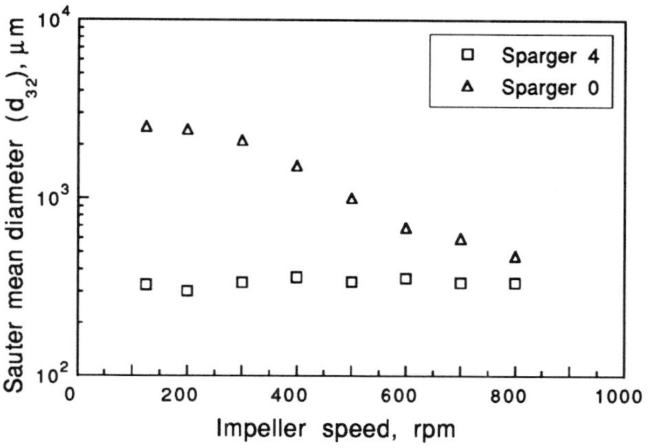

Figure 2. The effect of impeller speed on the Sauter mean bubble diameter for the Rushton turbine.

of sparger-0 bubbles for the Rushton turbine are shown in Figure 3 as a function of impeller speed. The bell shaped curve of the distribution on logarithmic axes at the initial impeller speed indicates that the distribution is log-normal (Figure 3a). The distribution curve was obtained by connecting the mid-points of the histogram bars representing the number frequency in each size band. As the impeller speed increases, the distribution becomes bimodal with the appearance of a second mode towards the lower side of the size scale at 200 rpm (Figure 3b). This would indicate the commencement of bubble break-up at this speed. With further increase in impeller speed, the size of the newly formed second mode gradually increases while that of the first one decreases (Figures 3c, 3d and 3e), because of further breakup of sparger generated bubbles. At even higher impeller speeds, the distribution reverts to the unimodal but with a slight skewness towards the larger

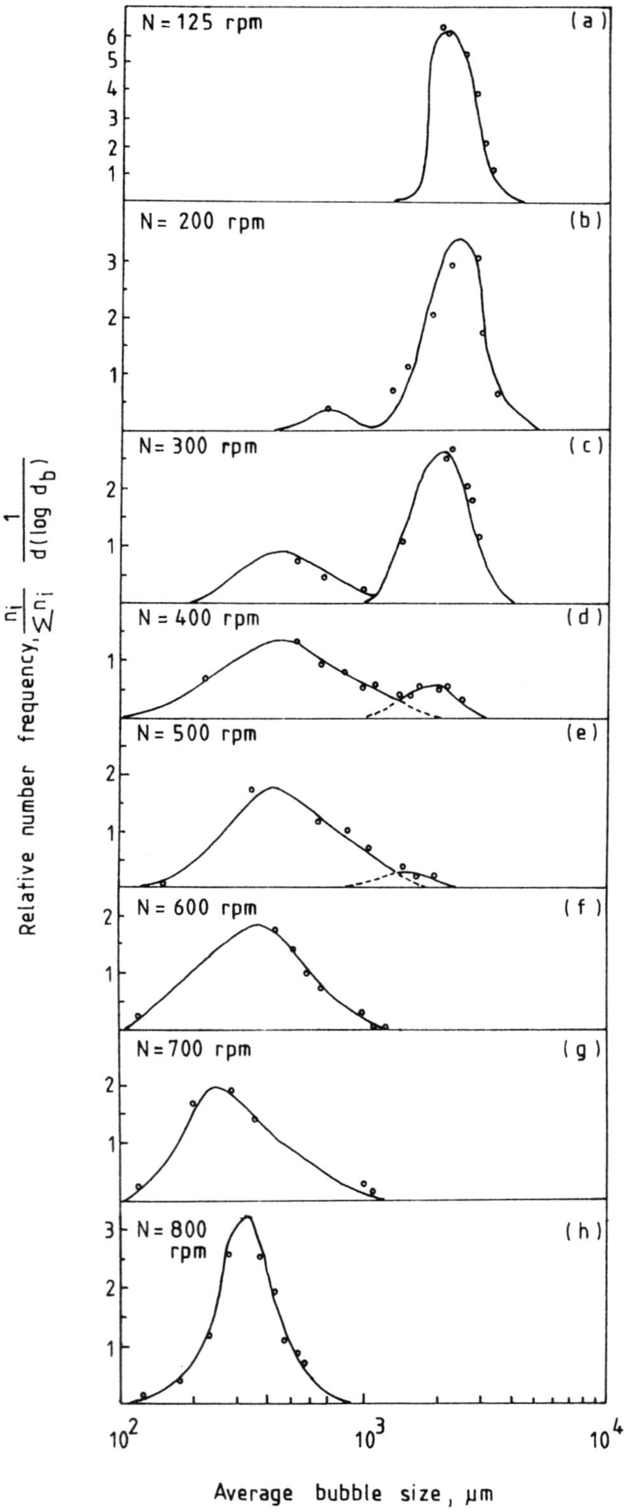

Figure 3. The effect of impeller speed on the frequency number distribution of bubbles generated by sparger-0 for the Rushton turbine.

bubble sizes (Figure 3f). This skewness disappears almost completely at 700 rpm (Figure 3g) and the bubble size distribution becomes unimodal log-normal again, but on the lower side of the bubble size scale (Figure 3h).

The first transition of unimodal to bimodal distribution occurs between 125 and 200 rpm and the next transition of bimodal to unimodal distributions takes place between 600 and 700 rpm. It is hard to pin-point a single impeller speed for the transitions as the changes in the size distributions take place over a range rather than at a particular impeller speed.

The above bubble size distribution data may also be represented as a cumulative (undersize) plot on log-normal probability graph as shown in Figure 4. The changes in the bubble size distribution over the range of impeller speeds studied can be recognized from this single graph. At the lowest impeller speed of 125 rpm, the plot is approximately linear which supports the observation that the distribution is log-normal. With increase in impeller speed, the distribution takes the shape of a characteristic curve as seen for 200 rpm. The curve can be viewed as a combination of two straight lines of different slopes representing two different bubble populations and thus a bimodal distribution. The linear dimensions of the straight lines represent the relative size fractions of the respective bubble populations. The percentage of bubbles under the mode on the smaller bubble size side is about 30% at this speed. At higher speeds, the distribution curve shifts to the left on the lower range of the bubble size scale. Also the segment of the curve representing the smaller bubbles increases in length indicating an increase in their number in the population, whereas the segment corresponding to the larger bubbles becomes smaller indicating their further breakup. This trend continues until 700 rpm when the distribution becomes linear again on the lower side of the scale. Thus, the linear-curved-linear trend of the

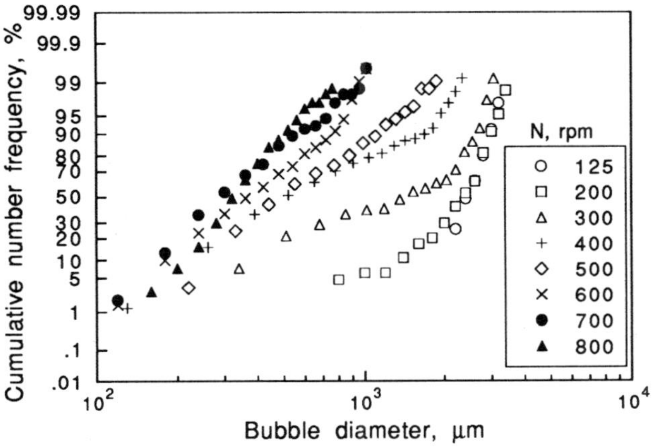

Figure 4. The effect of impeller speed on the cumulative size distribution (undersize) for sparger-0 bubbles, plotted on log-normal probability coordinates, for the Rushton turbine.

distribution curves on the cumulative plots is equivalent to the unimodal-bimodal-unimodal pattern of the distribution curves found on the frequency plots shown in Figure 3.

The pattern of unimodal-bimodal-unimodal change in the bubble size distribution with increase in impeller speed is observed for the pitched-blade disc turbine and the marine propeller also. The frequency number distribution of bubbles for these two impellers are shown in Figures 5 and 6, respectively, as a function of the impeller speed. Although the first unimodal-bimodal transition in the bubble size distribution is present for these two impellers, the subsequent bimodal-unimodal transition is present only for the pitched-blade disc turbine, within the range of impeller speed used. For the marine propeller, it does not occur even at the highest impeller speed used (900 rpm). A major proportion of bubbles in the population still remains in the first mode of the bimodal bubble size distribution at this speed. The cumulative size distribution plots for the above data are shown in Figures 7 and 8 for the pitched-blade disc turbine and the marine propeller respectively.

It can be seen from the above figures that the transitions in the form of bubble size distribution

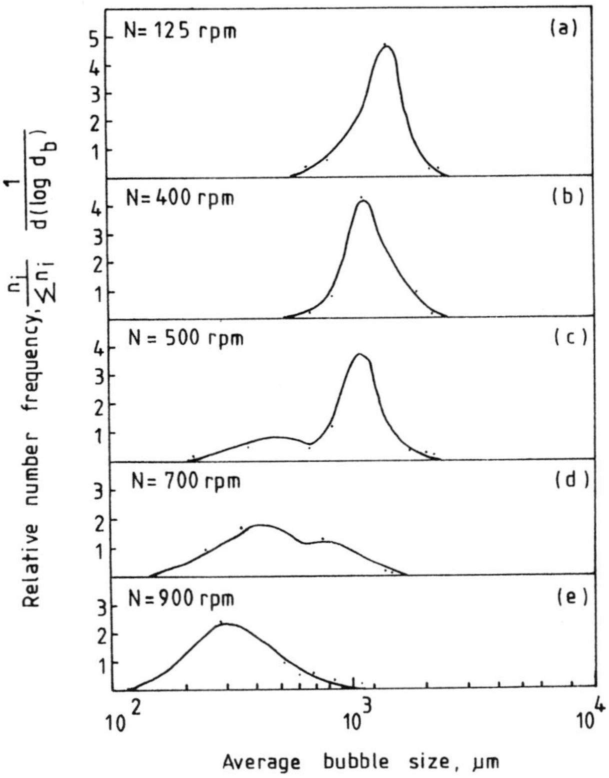

Figure 5. The effect of impeller speed on the frequency number distribution of bubbles generated by sparger-0 for the pitched blade impeller.

occur at different impeller speeds for the pitched-blade disc turbine and the marine propeller, when compared with those for the Rushton turbine. The first unimodal-bimodal transition in bubble size distribution occurs between 400 and 500 rpm for both the pitched-blade disc turbine and the marine propeller. The following bimodal-unimodal transition for the pitched-blade disc turbine occurs between 900 and 1000 rpm.

Sparger-4 bubbles. The number distribution of sparger-4 bubbles at the lowest impeller speed (125 rpm) was found to be well represented by log-normal distribution for all the three impellers used. This initial bubble size distribution was found to remain virtually unaffected with increase in impeller speed, as shown in Figure 9 for Rushton turbine. The mean bubble size given by the 50%

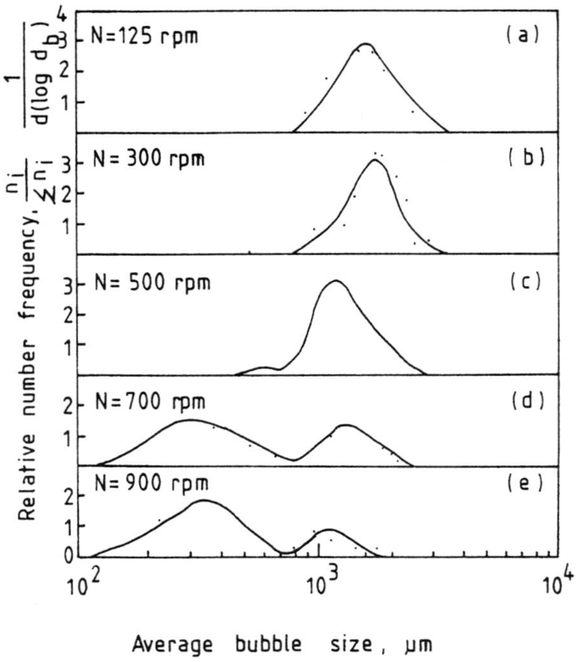

Figure 6. The effect of impeller speed on the frequency number distribution of bubbles generated by sparger-0 for the marine propeller.

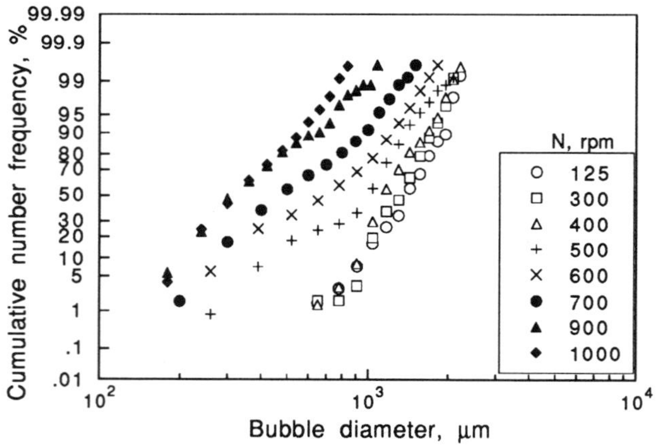

Figure 7. The effect of impeller speed on the cumulative (undersize) bubble size distribution for sparger-0 plotted on log-normal probability coordinates, for the pitched blade impeller.

value also does not change significantly with increase in impeller speed. These observations are consistent with the Sauter mean bubble diameter results shown in Figure 2. Similar results were obtained for the other two impellers.

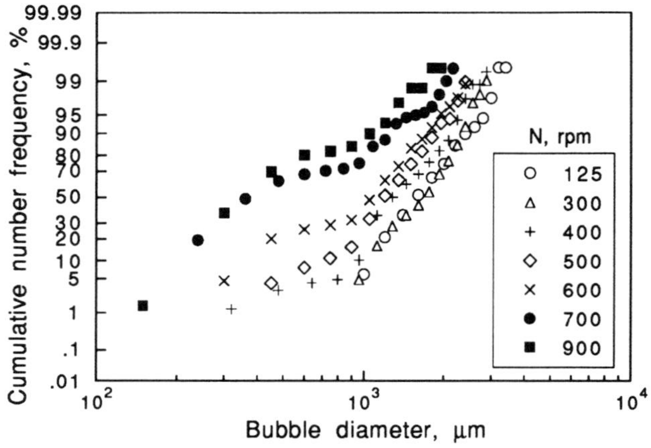

Figure 8. The effect of impeller speed on the cumulative bubble size distribution (undersize) for sparger-0 plotted on log-normal probability coordinates, for the marine propeller.

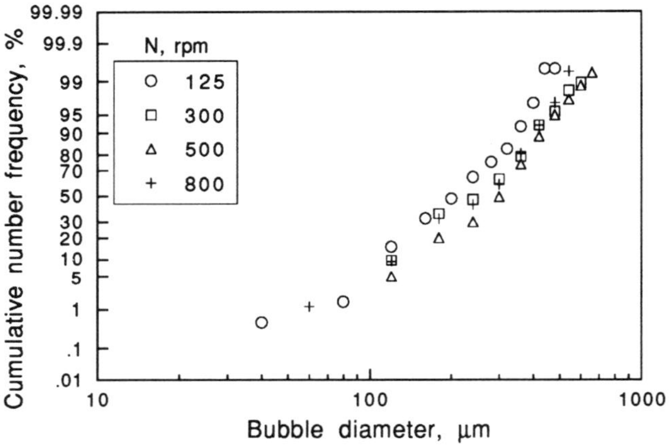

Figure 9. The effect of impeller speed on the size distribution of bubbles generated by sparger-4 plotted on log-normal probability coordinates, for the Rushton turbine.

The breakup behaviour of bubbles observed in this study could be explained by considering the concepts proposed by Hinze [8] for bubble/drop deformation in turbulent flow. For a given turbulence level, the bubbles larger than a certain maximum stable size (d_{max}) are expected to be broken up by the eddies whose length scale are similar to d_{max}. When the impeller speed is increased, the d_{max} value decreases, with the system reaching a new steady state at each impeller speed. The sparger-0 bubbles follow this pattern exactly. With commencement of agitation, bubbles larger than d_{max} breakup leading to the appearance of a new mode in the distribution in the direction of smaller bubbles. With further increase in agitation level, the size of the maximum stable bubble decreases and therefore more bubbles are broken up. As a consequence, the number of smaller bubbles in the population increases, as reflected in the progressively larger fraction in the second mode on the graph. With sufficient power input, as in the case of the Rushton turbine and the pitched blade impeller, most of the bubbles originally generated by the sparger would be expected to have broken up, and the resulting bubble size distributions reach new steady-state values, represented by unimodal distributions of smaller mean diameters.

The sparger-4 bubbles, being much smaller than the d_{max} value corresponding to even the highest level of agitation used in this study, remain unbroken and their average bubble size and bubble size distribution are preserved in the range of impeller speeds used.

It should be noted that the results presented above are for a non-coalescing environment. In coalescing media, the size distribution of bubbles undergo very rapid and significant changes as they move away from the impeller zone, which is the region for bubble breakup. Furthermore, the bubble size distributions measured in this study represent the steady-state bubble size distributions. The bubble size distribution results from the circulation of bubbles through the impeller (breakup) zone, and the number of circulation would be a function of the impeller type and speed and the gas flow rate. The bubbles going through the impeller zone at any moment of time would consist of bubbles entrained by the liquid streams and the primary bubbles produced in the impeller zone during their first passage through the impeller. The present state of knowledge on bubble circulation in a stirred vessel does not allow us to estimate the fraction of these two bubble groups in the measured bubble size distribution. Further work is required to determine the relationship between the size distributions observed and the power input into the system, or if possible, the number of passes an element of gas makes through the impeller during its stay in the dispersion. However, in the ultimate analysis it is likely that the observed effects could all be explained in terms of the energy dissipated in the system.

CONCLUSIONS

By generating bubbles of known initial sizes and, by using a non-coalescing system, it has been possible to observe the breakup of bubbles, as a function of agitation in an aerated stirred vessel. A number of impellers, incorporating radial, mixed and axial flow types, have been used.

Size distribution of bubbles with Sauter mean diameter of around 300 μm remain unaffected as they do not undergo any breakup with increase in impeller speed. Their size distribution remains log-normal for the range of impeller speed used in this study. On the other hand, bubbles of initial mean diameters of around 2.5 mm are broken up and their size distribution changes as a function of the impeller speed. The original log-normal distribution becomes bimodal due to the generation of smaller bubbles from the breakup of larger ones in the population. At higher impeller speeds, the bimodal distribution reverts to unimodal again, but on the lower side of the bubble size scale. This would indicate that the bubbles above a certain size are broken with each increment in agitation level, and attain a new steady-state distribution and thus a new mean size. The pattern of change in bubble size distribution from unimodal to bimodal to unimodal is influenced by the impeller type, indicating it is a function of specific power input into the system.

The reported observations, from carefully controlled experiments, provide a convenient backdrop for future theoretical deliberations in this area. The findings should be applicable to chemical and biochemical reactors, where the presence of salts or other surface active agents make the medium non-coalescing.

LITERATURE CITED

1. van't Riet, K. and J.M. Smith, Chem. Eng. Sci., **28**, 4, 1031 (1973).

2. Bruijn, W., K. van't Riet and J.M. Smith, Trans. Instn. Chem. Engrs., **52**, 1, 88 (1974).

3. Kawecki, W. T. Reith, T., J.W. van Heuven, and W.J. Beek, Chem. Eng. Sci., **22**, 11, 1519 (1967).

4. Barigou, M. and M. Greaves, Trans. I. Chem. E., Chem. Eng. Res. Des., **70**, A2, 153 (1992).

5. Barigou, M. and M. Greaves, Chem. Eng. Sci., **47**, 8, 2009 (1992).

6. Takahashi, K., W.J. McManamey and A.W. Nienow, J. Chem. Eng. Japan., **25**, 4, 427 (1992).

7. Parthasarathy, R., G.J. Jameson and N. Ahmed, Trans. I. Chem. E., Chem. Eng. Res. Des., **69**, A4, 295 (1991).

8. Hinze, J.O., 1955 A.I.Ch.E.J., **1**, 1, 289 (1955).

NOMENCLATURE

d_{32} Sauter mean bubble diameter, μm
d_{max} Maximum bubble diameter, μm
N impeller rotational speed, rpm
U_g superficial gas velocity, m/s

Minimun Agitation Speed for Off-Bottom Suspension of Solids in Agitated Vessels Provided with Multiple Flat-Blade Impellers

Piero M. Armenante and Tong Li

Dept. of Chem. Eng. Chem. and Environmental Science, New Jersey Inst. of Technology, Newark, NJ 07102

The effect of the presence of multiple flat-blade turbines on the attainment of the minimum agitation speed N_{js} to just suspend solid particles in agitated vessels was studied. The agitation system consisted of a stirred tank provided with one, two, or three impellers, and in which glass particles, 110 μm in diameter, were suspended. Water was used as the liquid. A strain gage system was used to determine the total power consumption, as well as the power dissipated by each individual impeller. It was found that the dependence of N_{js} from the impeller diameter for the case in which multiple impellers were used closely resembled that obtained when a single impeller was used. The role of the number of impellers, and of several geometric parameters, such as the clearance of the lower impeller off the tank bottom and the distance between impellers, was studied. In all cases, the value of N_{js} measured when multiple impellers were used was either nearly the same as that for the analogous single impeller case or higher. The corresponding power consumption was always higher than for single impellers. The lower impeller plays a major role in the off-bottom suspension of the solids, and the presence of additional flat-blade turbines either does not affect the suspension process or actually interferes with it. The presence of more than one flat-blade turbine is either unnecessary for, or actually detrimental to, the achievement of the solid suspension state. Therefore, their use is not justified if off-bottom solid suspension is the sole mixing requirement.

INTRODUCTION

The use of multiple impellers in mixing vessels, fermenters, and reactors is extremely common in the industrial practice. Nevertheless, the study of the mixing characteristics of vessels provided with such multiple agitation system has received much less attention than the case in which a single impeller is used. For example, if one examines the problem of suspending solid particles off the bottom of mixing vessels provided with a single impeller the literature on the subject is quite extensive (Baldi *et al.*, 1978 [3]; Barresi and Baldi, 1987 [4]; Chapman *et al.*, 1983 [5]; Narayanan *et al.*, 1969 [11], Nienow, 1968 [12]; Nienow, 1985 [13]; Zwietering, 1958 [14]). However, if multiple impellers are considered the available information is extremely limited. In recent work conducted by this group (Armenante *et al.*, 1992 [1]) significant similarities were found between the case of solid particle suspension and liquid-liquid dispersion in vessels provided with one or more impellers. In addition, the same work showed that in many situations the presence of multiple impellers reduced the minimum agitation speed requirement to produce the just suspended or just dispersed state, but the power consumption required for this purpose was significantly higher than that measured for single impellers. If this conclusion is generally valid, vessels provided with a single impeller would be more effective to just suspend solids than vessels provided with multiple impellers. In order to validate of disprove these conclusions the present investigation was carried out. The main objectives of this work were to determine the role of multiple impellers in the achievement of the just suspended state in mixing vessels provided with one or more flat-blade turbines, to determine the power consumption of the multiple impeller configuration and compare it with that for single impellers, and to quantify the effect of a number of geometric parameters such as off-bottom clearance, and impeller spacing.

APPARATUS AND PROCEDURE

A diagram of the equipment utilized in this work is shown in Figure 1. A 2 HP variable speed motor (G. K. Heller Corp.) was used to rotate a centrally mounted hollow shaft on which one, two, or three flat-blade turbine impellers could be mounted. The impellers had a diameter equal to 0.065, 0.076, or 0.102 m. Only impellers of the same size were use in any given experiment. The rotational speed was measured with an optical tachometer provided with a light sensor (Cole Parmer Instrument Co.), accurate within ±1 rpm. The torque applied to each impeller was measured by three strain gages mounted at different positions on the shaft. The gages were connected to a slip ring assembly (Electrominiature Corp.) with leads passing

* *Corresponding Author*

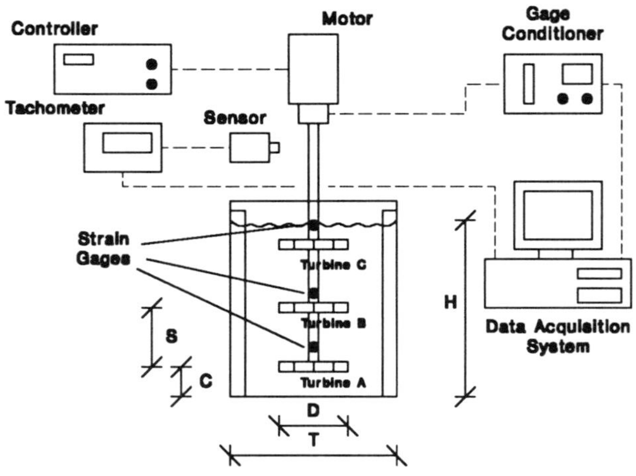

Figure 1. Schematic of the experimental apparatus used in this work

through the hollow shaft. The gage signal was picked up by an external gage conditioner and amplifier (Measurement Group, Inc., 2120 System). The signal was converted to torque using a calibration curve produced by applying known torques to the shaft under static conditions. A data acquisition system (Labtech Notebook) connected to a computer was used to collect the amplified signal on-line and to calculate the torque applied to each strain gage. The power consumed by each impeller was calculated from the product of the angular velocity and torque.

All experiments were carried out using with distilled water as a liquid in two Plexiglas tanks having a diameter equal to 0.241 and 0.292 m, respectively. The corresponding liquid heights were 0.305 and 0.330 m. The clearance of the lower impeller from the tank bottom, C, and the spacing between impellers, S, could be changed. Spherical glass particles having an average diameter of 110 μm (determined from the mesh size of the screen on which they were retained) served as the dispersed phase. The solid phase concentration in the liquid was always 0.5% by weight.

Each run always began at low agitation speed. At each agitation speed, the system was allowed approximately 10 to 15 minutes to reach steady state. The speed was increased after each observation until the point of just off-bottom suspension was reached. A mirror for the visual observation of the particle movements on the tank bottom was placed at a 45° angle under the vessel. Attention was paid to the location within the tank where the particles became suspended. This often provided information about the flow pattern near the tank bottom. The definition of minimum suspension speed, N_{js}, was taken as the speed at which no particles were visually observed to remain at rest on the tank bottom for more than one or two seconds (Zwietering, 1958 [14]). The reproducibility of the measurement was typically within ±3%, as previously reported (Armenante *et al.*, 1992 [1]).

For each minimum suspension speed, the corresponding power drawn by each impeller was experimentally obtained. The corresponding power number, N_p, was obtained from

$$N_p = P/(\rho_L N^3 D^5) \quad (1)$$

RESULTS AND DISCUSSION

The dependence of the minimum agitation speed

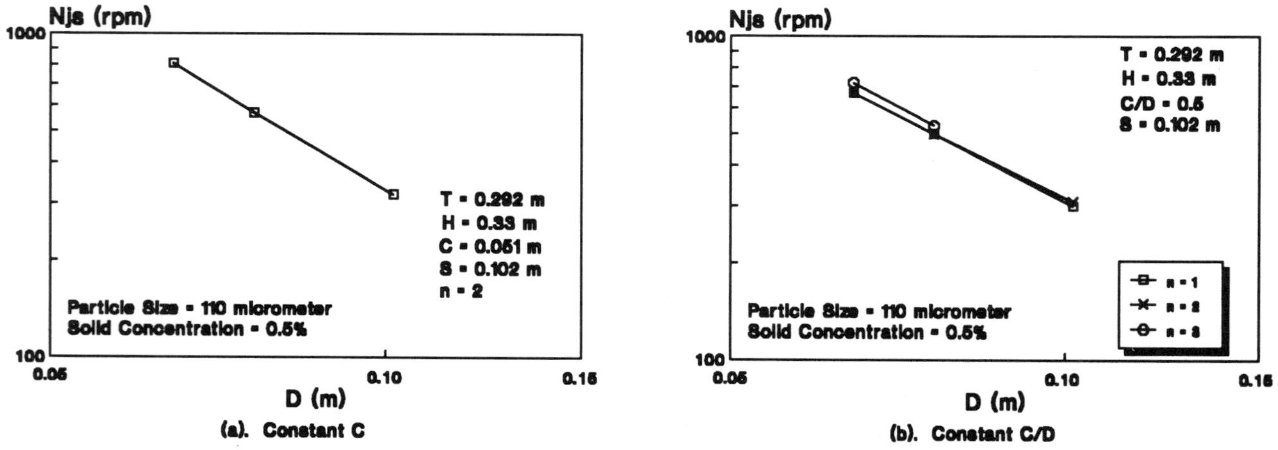

Figure 2. Effect of the impeller diameter, D, on N_{js}

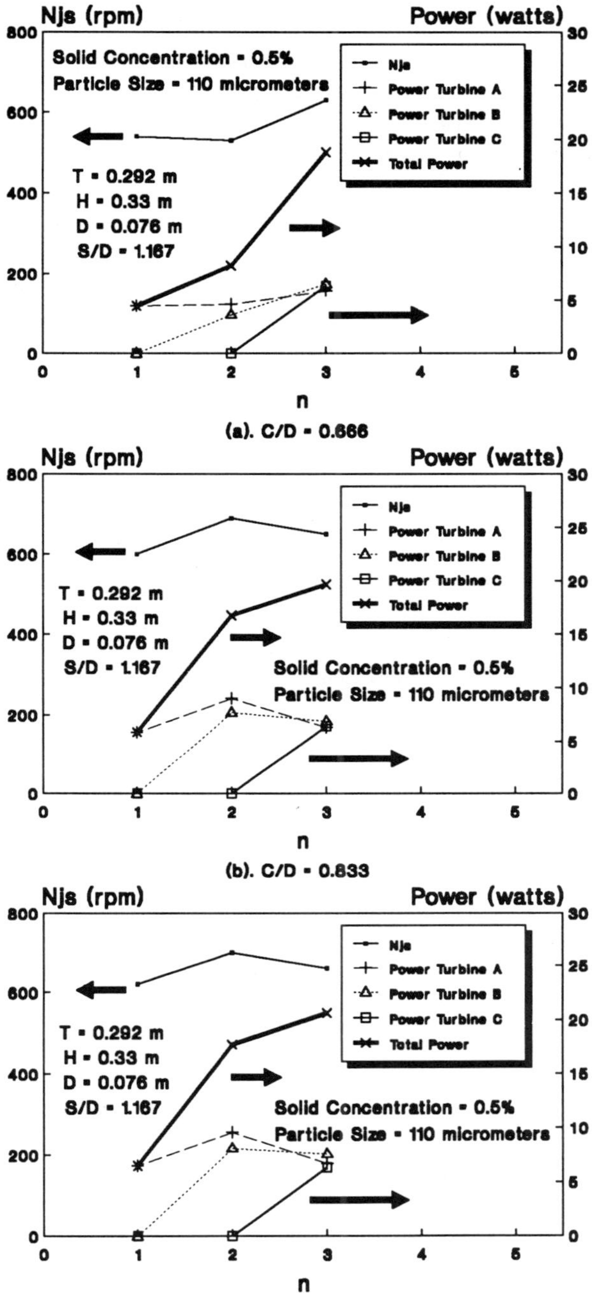

Figure 3. Effect of the number of impellers, n, on N_{js} for different C/D ratios

for off-bottom suspension using multiple impellers was obtained as a function of a number of geometric parameters. The effect on N_{js} of the impeller diameter, D, was determined for two cases, i.e., for constant clearance of the impeller off the tank bottom, and for constant C/D ratio. The results are shown in Figure 2a and Figure 2b, respectively. From Figure 2a an inverse relationship between N_{js} and D can be observed when two impellers are mounted on the same shaft. This dependence can be expressed as

$$N_{js} \propto D^{-2.05} \qquad (2)$$

where the standard error of the exponent is ±3.5%. The value for this exponent is very similar to those obtained in previous investigations for the case in which a single flat-blade turbine or a paddle was used, and the T/C ratio was kept constant. For example, the results of Zwietering (1958 [14]) for two types of paddles (having impeller diameter to blade height ratios equal to 2 or 4, respectively) can be shown to be well represented by an equation similar to Equation (1) in which the value of the exponent is 2.06 and 2.08, respectively. Similarly, the results obtained by Narayanan *et al.* (1969) [11] for flat-blade turbines with eight blades show that their value for the exponent of D in Equation (2) would be 2.0. The similarity between the results obtained here for two impellers and those obtained by those investigators for a single impeller indicate that the scale up factor for the two systems is the same. This can be interpreted as an indication that the lower impeller, which is closer to the solids being suspended, plays a key role in the suspension process.

For the case in which C/D is constant Figure 2b shows that a linear relationship also exists but that the dependence between N_{js} and D is now given by

$$N_{js} \propto D^{-1.77} \quad \text{for } n = 1 \qquad (3)$$

$$N_{js} \propto D^{-1.69} \quad \text{for } n = 2 \qquad (4)$$

$$N_{js} \propto D^{-1.95} \quad \text{for } n = 3 \qquad (5)$$

where the standard error of the exponents in Equations 3 and 4 is ±1.9% and ±3.5%, respectively. The values of all these exponents are similar to each other, and are smaller than that obtained for the case of constant C. A similar difference in exponent between the case in which C/D is a constant versus that in which C is constant has also been reported in the literature not only for the case of solid suspension but also liquid-liquid dispersion (Armenante *et al.*, 1992 [1]) and floating solid drawdown in liquids (Armenante *et al.*, 1991 [2]). A review of the literature on solid suspension also reveals that those investigators that quantified the effect of D on N_{js} at constant C (Narayanan *et al.*, 1969 [11]; Nienow, 1968 [12];

Zwietering, 1958 [14]) reported values for the exponent for D significantly smaller than those who kept C/D constant (Armenante, *et al.*, 1992 [1]; Armenante *et al.*, 1991 [2]; Baldi *et al.*, 1978 [3]). This conclusion appears to be independent of the type of radial impeller used.

The values for the exponents in Equation 3, 4, and 5 are also close to the theoretical value of -1.67 that was obtained by Baldi *et al.* (1978) [3] for solid suspension with a single impeller, using Kolmogoroff's theory of isotropic turbulence. Figure 2b shows that no appreciable difference exists between the values for N_{js}, independently of the number of impellers used (if anything, an increase for N_{js} can be observed when a third impeller is added.) This fact confirms that the presence of multiple impellers on the same shaft does not necessarily result in an improvement of the mixing performance, at least as far as the achievement of the off-bottom suspension state is concerned (Armenante *et al.*, 1992 [1]).

A similar conclusion can be drawn by examining the dependence of N_{js} on the number of impellers, n, as indicated in Figure 3. This figure shows that, independently of the value of the C/D ratio, N_{js} either remains constant or goes up as n increases. This implies that the presence of extra impellers either does not have any impact in promoting off-bottom suspension, or actually interferes with the solid suspension process. Variations in the flow pattern resulting from the presence of impellers have been recently described by a number of investigators (Chiampo *et al.*, 1991 [6]; Hudcova *et al.*, 1989 [9]; Mahmoudi and Yianneskis, 1991 [10]). In any case, the additional impellers are clearly associated with extra power consumption, as shown in Figure 3, even if the value of N_{js} does not change as n increases. Figure 3 also reports the power dissipated by each individual impeller. For the case in which only two impellers are present the lower impeller always consumes more power in correspondence of N_{js}, typically about 20% more then the top one. However, when three impellers are mounted on the same shaft the power consumption of the middle impeller is typically higher than the other two by about 10%. The conclusion that can be drawn from the results presented in Figure 3 is that there is no advantage in placing more than one flat-blade turbine on the same impeller if the objective is just to suspend solid particles off the tank bottom.

Figure 4 shows the effect of increasing the clearance off the tank bottom on N_{js}. From Figure 4a one can see that a near linear relationship exists between N_{js} and the ratio C/D for the case in which one impeller is used. However, the same does not apply when two impellers are present. As indicated in Figure 4b, when C/D increases the minimum agitation speed increases more than linearly, and so does the power. This figure also shows that the lower impeller consumes slightly less power than the top one, but that this difference disappears as C/D increases. For the case of one impeller it was noticed that the flow pattern below the impeller was a function of the impeller clearance off the tank bottom. For low C/D values (typically below 0.666, corresponding to C/T < 0.21) the flow pattern was observed to be swirling outwards so that the particle suspension occurred from the periphery of the tank. For high values of C/D (typically above 0.833, corresponding to C/T > 0.26) the swirling action turned inward, and solid suspension occurred

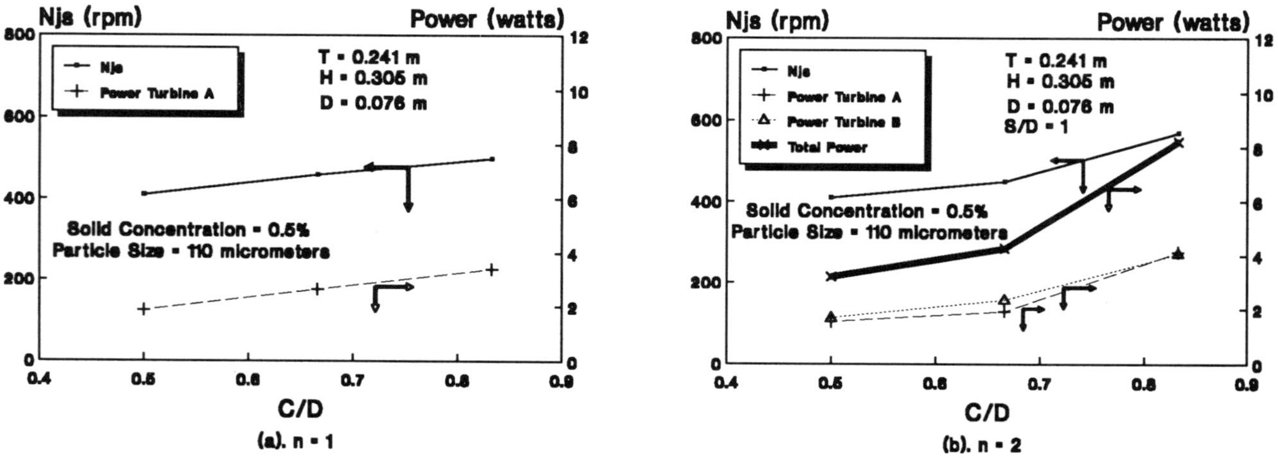

Figure 4. Effect of the C/D ratio on N_{js} for different agitation systems

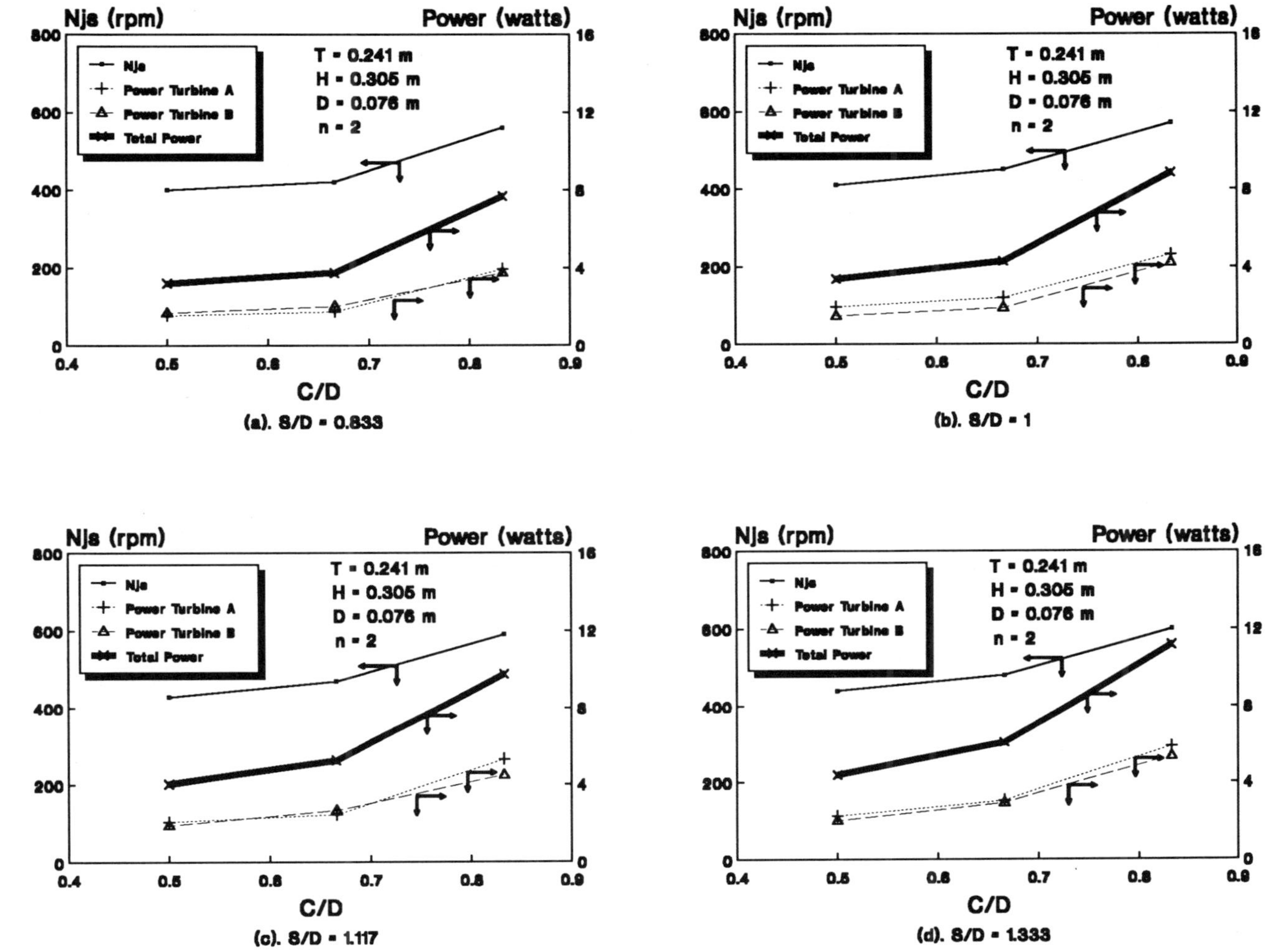

Figure 5. Effect of the C/D ratio on N_{js} for different S/D ratios

from the center of the tank bottom. For intermediate values of C/D a transition region consisting of an unstable flow pattern was noticed. The presence of different suspension regimes (from the tank periphery or center) is consistent with what has been observed by previous investigators in similar agitation systems. For example, Nienow (1968) [12] observed a similar change in flow pattern using a disk turbine when C/T was about 0.2. Conti *et al.* (1981) [7] reported that in correspondence of C/T = 0.22 the suspension mode switched from one regime to the other for a single 8-bladed disk turbine agitation system. Similarly, Gray (1987) [8] noticed that the flow pattern changed in correspondence of C/T = 0.17 for the case of a single flat-blade turbine. However, in the present work it was noticed that the range in which the presence of the transition region occurred (0.21 < C/T < 0.26) was independent of the presence of additional impellers, at least if the distance between impellers was kept equal to the impeller diameter. This again confirms that the lower impeller is primarily, if not solely, responsible for off-bottom solid suspension.

The dependence of N_{js} on the C/D ratio is also reported in Figure 5 for different values of the distance between the impellers (expressed here in terms of the ratio S/D). This figure shows that higher N_{js} values are required for increasing values of C/D, and that this increase is non-linear. This result is similar to that reported in other studies with single impellers (Conti *et al.*, 1981 [7]; Gray, 1987 [8]; Zwietering, 1958 [14]). In addition, this figure shows that as the distance between impeller increases the required agitation speed also increases, at constant C/D. This indicates that a closer proximity of the two impellers helps somewhat the lower impeller in its particle suspension action. However, this still does not justify the presence of the second impeller with respect to the case in which only one impeller is used.

Finally, Figure 6 gives the dependence of N_{js} from the distance between impellers, S (expressed in terms of the S/D ratio) at various clearances of the lower impeller off the tank bottom (expressed as C/D). One can clearly see from this figure that N_{js} is only a weak function of S. This figures also seems to indicate that for large values of the S/D ratio the power consumption reaches a plateau or even decreases.

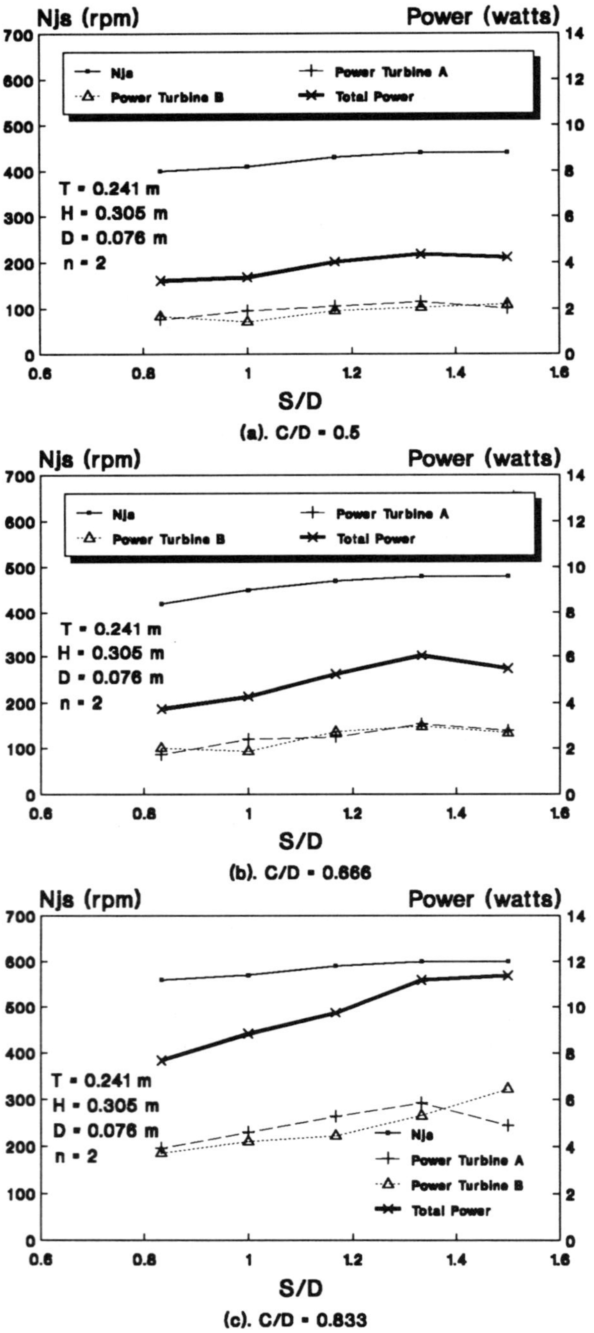

Figure 6. Effect of the S/D ratio on N_{js} for different C/D ratios

CONCLUSIONS

The results of this study indicate that the presence of more than one flat-blade turbine on the same shaft does not improve the performance of the agitation system, as far as the achievement of the just suspended state of finely divided solids is concerned. In the vast majority of cases it was noticed that the minimum agitation speed actually *increased* when more than one impeller were used. Even in those cases in which the value of N_{js} was independent of the number of impellers the power consumption was always higher than when a single impeller was used. These conclusions support the idea that the phenomenon of solid suspension off the tank bottom is largely dominated by the lower impeller, and that any interferences in the flow pattern generated by this impeller, such as those resulting from the presence of more impellers, can result in higher energy requirements to achieve the same result.

ACKNOWLEDGMENT

This work was partially supported by the Exxon Education Foundation (thanks to the interest of Ramesh Hemrajani of Exxon Research and Engineering) whose contribution is gratefully acknowledged.

NOTATION

C Clearance of bottom impeller off the tank bottom; m

C_t Distance of top impeller from air-liquid interface; m

D Impeller diameter; m

H Height of liquid in mixing vessel; m

n Number of impellers mounted on the same shaft, non-dimensional

N_p Power number, non-dimensional

N_{js} Minimum agitation speed for off-bottom suspension of solid particles in solid-liquid systems; revolutions per minute (rpm)

P Power consumption, watts

S Distance between impellers; m

T Vessel diameter; m

Greek Symbols

ρ_L Liquid density, m

LITERATURE CITED

1. Armenante, P. M., Huang, Y. T., and Li, T., Determination of the Minimum Agitation Speed to Attain the Just Dispersed State in Solid-Liquid and Liquid-Liquid Reactors Provided with Multiple Impellers, *Chem. Eng. Sci.*, **47**: 2865-2870 (1992).
2. Armenante, P. M., Mmbaga, J. P., and Hemrajani, R. R., Mechanisms for the Entrainment of Floating Particles in Mechanically Agitated Liquids, *Proc. 7th Europ. Conf. on Mixing, Brugge, Belgium, 18-20 Sept., 1991*, Vol. 2,: 555-564 (1991).
3. Baldi, G., Conti, R., and Alaria, E., Complete Suspension of Particles in Mechanically Agitated Vessels, *Chem. Eng. Sci.*, **33**: 21-25 (1978).
4. Barresi, A. and Baldi, G., Solid Dispersion in an Agitated Vessel, *Chem. Eng. Sci.*, **42**: 2949-2956 (1987).
5. Chapman, C. M., Nienow, A. W., Cooke M. and Middleton, J. C., Particle-gas-liquid mixing in stirred vessels. Part 1: particle-liquid mixing, *Chem. Eng. Res. Des.*, **61**: 71-81 (1983).
6. Chiampo, F., Guglielmetti, R., Manna, L., and Conti, R., Gas-Liquid Mixing in a Multiple Impeller Stirred Vessel, *Proc. 7th Europ. Conf. on Mixing, Brugge, Belgium, 18-20 Sept., 1991*, Vol. 2,: 333-341 (1991).
7. Conti, R., Sicardi, S, and Specchia, V., Effect of the Stirrer Clearance on Particle Suspension in Agitated Vessels, *Chem. Eng. J.*, **22**: 247-249 (1981).
8. Gray, D. J., Impeller Clearance Effect on Off Bottom Particle Suspension in Agitated Vessels, *Chem. Eng. Comm.*, **61**: 151-158 (1987).
9. Hudcova, V., Machon, V. and Nienow, A. W., Gas-Liquid Dispersion with Dual Rushton Turbine Impellers, *Biotech. Bioeng.*, **34**: 617-628 (1989).
10. Mahmoudi, S. M. and Yianneskis, M, The Variation of Flow Pattern and Mixing Time with Impeller Spacing in Stirred Vessels with Two Rushton Impellers, *Proc. 7th Europ. Conf. on Mixing, Brugge, Belgium, 18-20 Sept., 1991*, Vol. 1: 17-24 (1991).
11. Narayanan, S., Bhatia, V. K., Guha, D. K., and Rao, M. N., Suspension of Solids by Mechanical Agitation, *Chem. Eng. Sci.*, **24**: 223-230 (1969).
12. Nienow, A. W., Suspension of solid particles in turbine agitated baffled vessels, *Chem. Eng. Sci.*, **23**: 1453-1459 (1968).
13. Nienow, A. W., The Dispersion of Solid in Liquids, in *"Mixing of Liquids by Mechanical Agitation,"* by J. J. Ulbrecht and G. K. Patterson (editors), p. 273-307, Gordon and Breach Science Publishers (1985).
14. Zwietering, T. N., Suspending of solid particles in liquids by agitators, *Chem. Eng. Sci.*, **8**: 244-253 (1958).

Comminution Phenomena in Stirred Sugar Suspensions

Barbara Mazzarotta

Dept. of Chemical Engineering, University of Rome "La Sapienza"
via Eudossiana 18, 1-00184, Rome, Italy.

Size reduction phenomena occurring in sugar suspensions were investigated to determine the final crystal size distributions (CSDs), to ascertain the prevailing fracture mechanism (abrasion or shatter) and to evaluate the influence of suspension density, stirring rate and vessel geometry.
Large regular sugar crystals (1.4-1.18 mm) were suspended in a non-solvent liquid (xylene) inside a flat bottomed vessel fitted with 4 baffles and a central draft-tube: a spread of fragments was generated, the fines (<0.425 mm) being less than 0.5 wt. %. Comminution phenomena increased with suspension density and stirring rate, while slightly decreased by removing the baffles and the draft-tube.
The CSDs were satisfactorily fitted by a simple model which combines abrasion and shatter fracture and contains three adjustable parameters. Abrasion and shatter fracture coexisted, although the first mechanism predominated.

Comminution of the solid particles is a rather undesirable consequence of stirring solid-liquid suspensions, mainly due to impeller-particle, wall-particle, and particle-particle collisions. Size reduction is more evident using fragile particles, as those involved in fermentation and crystallization processes. In the latter case comminution phenomena greatly contribute to secondary nucleation [1,2].

The fracture of the particles may take place according to abrasion or shatter mechanisms, mainly depending on the impact energy and on the fracture strength of the particles [3]. Abrasion occurs when the impact energy is just sufficient to fracture a small volume of the particle: very fine fragments are generated and a coarse product, comparable in size with the initial particle, remains. Breakage takes place when the impact energy well exceeds the fracture strength of the particle, which is completely shattered, generating fragments spreading in a wide size range. Therefore, it is possible to discriminate abrasion and shatter fracture also basing on the size distribution of the generated fragments: in the first case two distinct regions of coarse and fine fragments are observed, with a high peak in the first one and a very low peak in the second; in the latter case the fragments range from coarse to fine with a single peak at an intermediate size. Figure 1 shows the typical cumulative wt. size distributions of the product obtained from abrasion and shatter fracture, respectively.

Size reduction occurring to potassium sulphate crystals suspended in a hydro-alcoholic saturated solution was investigated in a recent paper [4]. Due to the tendency of this salt toward agglomeration [5] a very long preliminary recrystallization procedure was needed to ensure to each set of the used crystals reproducible characteristics. The crystals underwent sensible comminution, which increased on increasing time, stirring rate and supersaturation used during the recrystallization, while it appeared not very sensitive to variations in the suspension density, in the investigated range (13-40 kg/m^3). The relative influence of abrasion and shatter fracture was highlighted, showing that shatter played a minor, but not negligible, role.

The aim of the present work is to extend the investigation on size reduction phenomena occurring in crystalline systems. Reference is made to quite concentrated suspensions of sugar crystals, in order to characterize their fracture mechanisms and to evaluate the influence of some operating variables (suspension density, stirring rate and vessel geometry) on comminution phenomena.

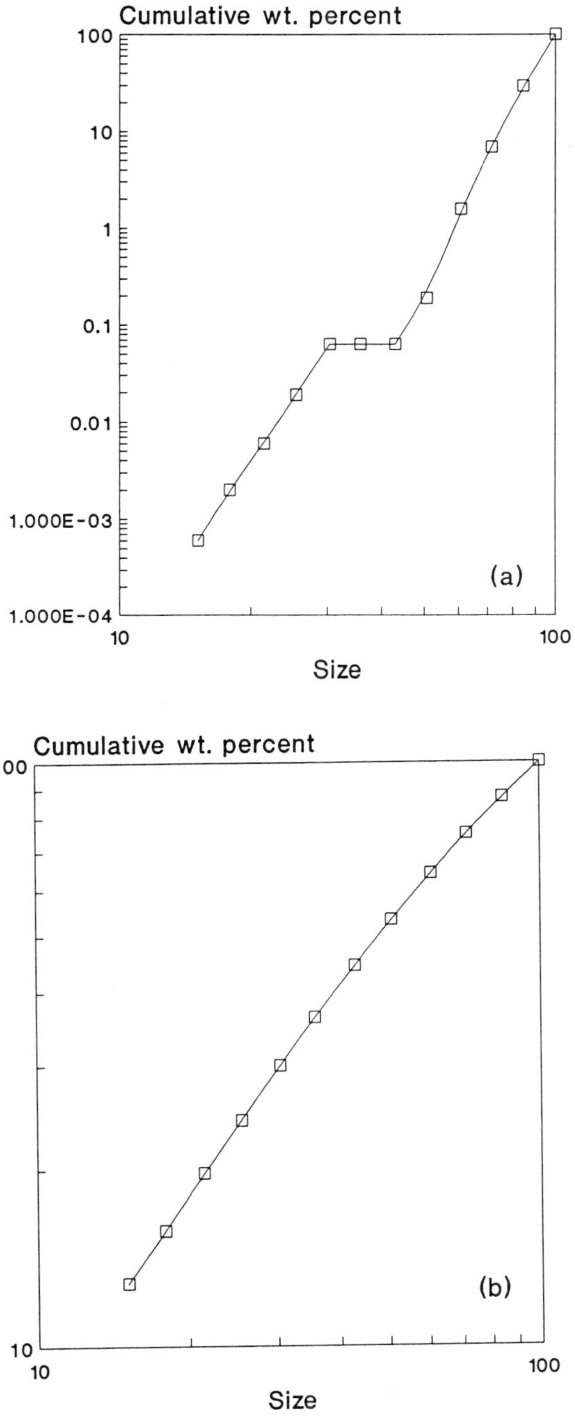

Figure 1. Typical Cumulative Wt. Size Distribution of Fragments Generated by Abrasion (a) and Shatter (b) Fracture Mechanism

EXPERIMENTAL

The experiments were carried out on sugar crystals (density 1590 kg/m^3) suspended in xylene (density 860 kg/m^3 and viscosity 0.65 mPa s).

The used crystals were commercial product obtained from sugar-cane which presented a very regular crystal habit, making the recrystallization step unnecessary. The crystals were preliminary dried and then sieved, selecting the sieve fraction 1.4-1.18 mm. Xylene was choosen as the suspending liquid since sugar is completely insoluble into this compound, thus avoiding that any crystallization phenomena, such as generation of nuclei or dissolution, could be superimposed to comminution due to mechanical fracture.

The experimental apparatus was a glass cylindrical flat-bottomed vessel (diameter 200 mm, height 350 mm) stirred by a three-blade marine-type propeller (diameter 70 mm) located at 50 mm above the bottom, and fitted with a central draft-tube (diameter 140 mm) and 4 radial baffles (width 20 mm). The group formed by the draft-tube and by the baffles (assembled in plexiglass) was easily removable from the vessel.

The runs consisted of stirring different suspensions, prepared by adding the desired amount of sugar crystals to 5 L of xylene, for a fixed time of 1 h, then determining the final size distribution of the suspended crystals. To this end the whole suspension was carefully discharged and immediately filtered under vacuum: the recovered crystals were dried and sieved by using standard ASTM series of sieves, down to 0.212 mm.

All the runs were carried out at room temperature, which resulted in a temperature of the suspension in the range 18-20°C. The stirring rate was constant within 2 rpm. The product lost in discharging and filtering the suspension, and in sieving the recovered crystals, was less than 0.5 wt.%.

Three series of runs were carried out by varying the suspension density, the stirring rate and the geometry of the vessel, with and without the group formed by the draft-tube and the baffles (indicated as DT+B), as listed in Table 1.

RESULTS

Some reproducibility tests were carried out: the maximum difference of the values of each sieve fraction was within 10 %.

The final crystal size distributions (CSDs) obtained from the experimental runs were plotted as cumulative weight percent, W, vs. size, L, in log-log graphs, as shown in Figure 2.

The trend of these curves seems quite similar to that typical of abrasion fragments (see Figure 1.a). In fact: (i) a very coarse product is present; (ii) the slope of the curves decreases drastically as size decreases, meaning a strong reduction of the amount of fragments; (iii) particularly at low suspension density, the slope of the curves slightly increases in the zone of fine fragments. The presence of little shatter fracture is evidentiated by the slight slope exhibited by the curves in the intermediate zone.

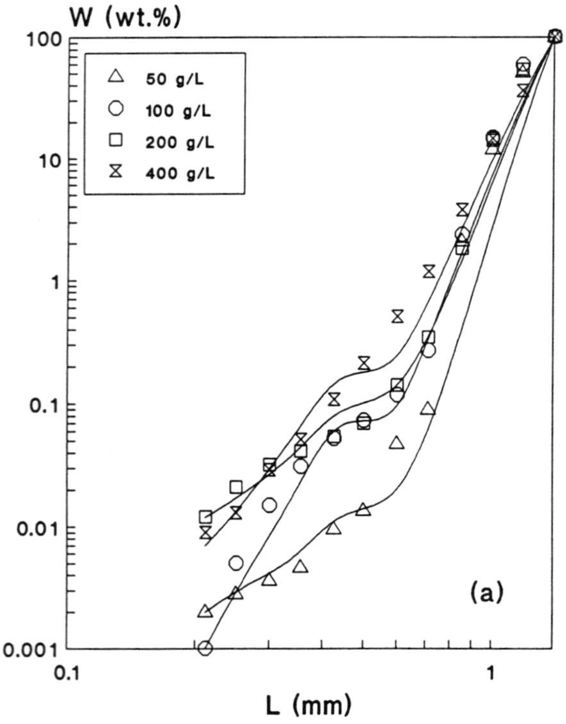

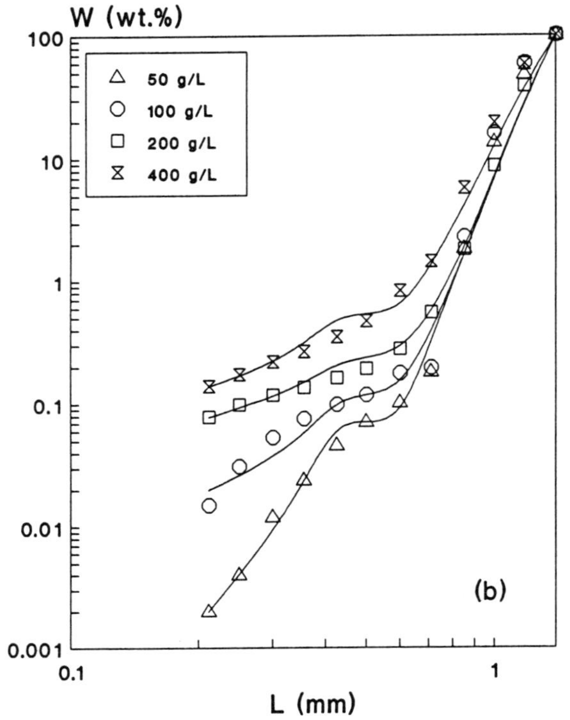

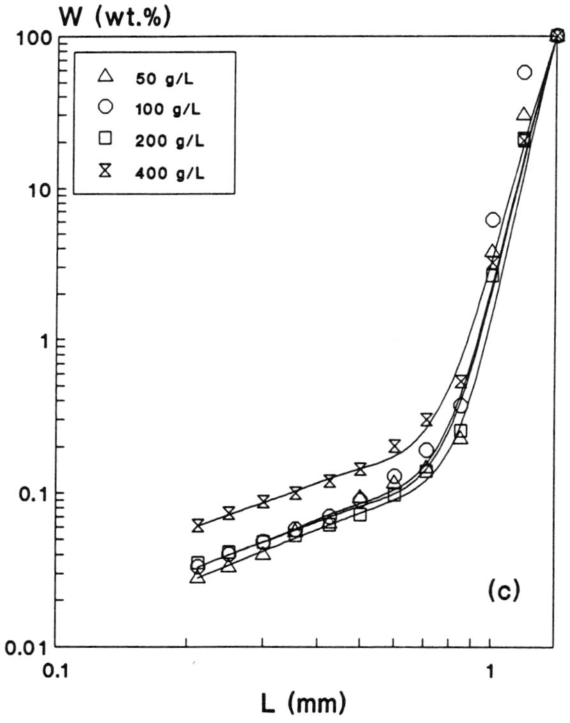

Figure 2. Experimental and Calculated CSD's: a) 800 rpm With Draft-Tube and Baffles; b) 1000 rpm With Draft-Tube and Baffles; c) 1000 rpm Without Draft-Tube and Baffles

Comminution phenomena increased, i.e. the curves shifted toward higher W values, on increasing the suspension density and the stirring rate. The influence of the suspension density is more evident at high stirring rate, probably due to the better mixing conditions achieved at 1000 rpm; the value of 800 rpm, in fact, roughly corresponds to the stirring rate ensuring complete off-bottom suspension at a suspension density of 100 kg/m^3.

The strong change in the fluid dynamics in the vessel, obtained by removing the group formed by the draft-tube and by the baffles, resulted in CSDs quite insensitive to variations of the suspension density: a slight increase of size reduction phenomena is noticed only at the highest suspension density value (400 kg/m^3). Moreover, comminution is greater in the modified than in the original geometry at low suspension density, but it is smaller at high suspension density.

MATHEMATICAL SIMULATION

A simple abrasion + shatter model, described in details elsewhere [4] was used to fit the CSDs.

The model assumes that the cumulative weight fraction of the fragments, y_w, can be calculated by linearly combining the cumulative wt. fractions of the fragments separately generated by abrasion and shatter fracture, respectively $y_{w,a}$ and $y_{w,s}$:

$$y_w = (1-k_s)\, y_{w,a} + k_s\, y_{w,s} \qquad (1)$$

where k_s is the fraction of fracture due to shatter with respect to the overall fracture (and $1-k_s$ is that due to abrasion).

$y_{w,a}$ and $y_{w,s}$ are functions of the dimensionless size $x = L/L_{max}$.

Abrasion fragments are assumed to be either coarse ($x > x_c$) or fine ($x < x_f$), with x_c and x_f minimum size of the coarse fragments and maximum size of the fine ones, respectively.

Any usual particle size distribution expression available in the literature can be adopted to express the functions $y_w(x)$; by using the Broadbent and Callcott relationship [6] it can be written:

$$y_{w,s} = \frac{1 - \exp(-x)^{n_s}}{1 - \exp(-1)} \qquad 0 < x < 1 \qquad (2)$$

$$y_{w,a} = \frac{1 - \exp(-x)^{n_c}}{1 - \exp(-1)} \qquad x_c < x < 1 \qquad (3a)$$

$$y_{w,a} = \frac{1 - \exp(-x_c)^{n_c}}{1 - \exp(-1)} = \frac{1 - \exp(-x_f)^{n_f}}{1 - \exp(-1)} \qquad x_f < x < x_c \qquad (3b)$$

$$y_{w,a} = \frac{1 - \exp(-x)^{n_f}}{1 - \exp(-1)} \qquad 0 < x < x_f \qquad (3c)$$

It appears also convenient to introduce y_{wf}, cumulative weight fraction of fine fragments (smaller than x_f), generated either by abrasion or by shatter, which can be directly derived from Equations (1,2,3c):

$$y_{wf} = (1-k_s)\frac{1 - \exp(-x_f)^{n_f}}{1 - \exp(-1)} + k_s\frac{1 - \exp(-x_f)^{n_s}}{1 - \exp(-1)} \qquad (4)$$

In conclusion, the model consists of Equations (1-4) and contains five independent parameters: x_f, x_c, y_{wf}, k_s, n_s.

The model was applied to fit the present CSDs. The values of the parameters x_f and x_c were evaluated reading on the graphs the L values at which the slopes of the curves changed: the values $x_f = 0.3036$ and $x_c = 0.4286$ were taken as constant for all the runs.

The values of the remaining three parameters, y_{wf}, k_s and n_s, were calculated for each run by applying a regression procedure and minimizing the percent deviations between the calculated and the experimental values of y_w. The obtained results are listed in Table 1.

Table 1. Operating Conditions and Calculated Values of the Parameters

Run No.	c kg/m^3	n 1/s	DT+B	y_{wf}	n_s	k_s
1A	50	13.3	yes	0.0001	1.84	0.0004
1B	100	13.3	yes	0.0007	2.60	0.0004
1C	200	13.3	yes	0.0009	1.95	0.0030
1D	400	13.3	yes	0.0016	2.58	0.0040
2A	50	16.7	yes	0.0007	1.38	0.0001
2B	100	16.7	yes	0.0011	1.40	0.0018
2C	200	16.7	yes	0.0021	1.07	0.0039
2D	400	16.7	yes	0.0051	0.88	0.0049
3A	50	16.7	no	0.0006	1.23	0.0019
3B	100	16.7	no	0.0007	1.18	0.0020
3C	200	16.7	no	0.0007	1.12	0.0019
3D	400	16.7	no	0.0013	1.05	0.0030

The calculated curves fitted quite well the experimental CSDs, as shown in Figure 2.

DISCUSSION

In solid-liquid suspensions comminution takes place according to the energy and the frequency of the impacts of the particles with the impeller, with the walls and the other fittings of the vessel, and with the other particles. In the first two cases the impact energy is proportional to the square of the impeller speed and the collision frequency is proportional to the impeller speed and to the reciprocal of the suspension volume; in the last case the impact energy is proportional to the square of the particle-fluid relative velocity, and the collision frequency is proportional to the concentration of solids and to the particle-fluid relative velocity [7].

Therefore, size reduction phenomena are expected to increase by increasing both the stirring rate and the suspension density, the latter mainly affecting particle-particle collisions.

The present results confirm the above mentioned influence of the stirring rate and of the suspension density, which is also shown by the values of the parameters y_{wf} and k_s (see Table 1). They both increase mainly on increasing the suspension density, and also on increasing the stirring rate: the influence of the suspension density appears remarkably higher when the vessel is fitted with the baffles and the draft-tube, due to the different fluid dynamics in the two cases. In fact, when the baffles are removed, the suspension is not uniformly placed in the vessel, but accumulates in the annular region along the wall.

Large part of the observed size reduction seems to depend on particle-particle collisions, particularly when the draft-tube and the baffles are present, although impeller-particle and wall-particle collisions are important too, especially at low stirring rate, i.e. when good mixing conditions are not completely developed.

The total amount of fine fragments ($L < 0.425$ mm) is very low, varying from 0.01 to 0.51 wt. %. In a previous work [4] potassium sulphate crystals sized 1-1.18 mm were submitted to similar experiments at a suspension density equal to 26.7 kg/m^3, obtaining a total amount of 0.1-0.15 wt. % fine fragments ($L < 0.125$ mm). These values are considerably higher than the present ones (0.01-0.07 wt. % at 50 kg/m^3), also taking into account that the fragment size is much smaller. Therefore, the characteristics of the crystalline system are confirmed [4] to strongly influence comminution phenomena: the regular habit of sugar crystals gave them a much higher resistance with respect to the partially agglomerated [5] potassium sulphate crystals.

Abrasion is confirmed to be the most effective fracture mechanism in crystalline solid-liquid suspensions [2]. In the present case, in particular, the percentage of crystal fracture due to shatter is very low, ranging from 0.01 to about 0.5 %, values lower than those (1.9-2 %) obtained for potassium sulphate crystals in the previously cited work [4]. This difference arises in part from the different characteristics of the crystals and in part from the adopted operating conditions: in fact, partially agglomerated crystals, such as the potassium sulphate ones, can be shattered more easily than the sugar ones, and abrasion is more likely to occur at high suspension density values, such as those adopted in the present work.

The modulus of the particle size distribution relevant to shatter mechanism, n_s, decreased by increasing the stirring rate, while it appeared not much affected by the other variables. The values obtained in the present work are quite similar to those previously derived [4] for potassium sulphate crystals, so that the shape of the curves relevant to the fragments generated by shatter fracture does not appear to be substantially modified by changes in the crystalline system.

CONCLUSION

Comminution phenomena in agitated sugar suspensions increase by increasing the suspension density and the stirring rate, but seem to be quite insensitive to these variables when both the baffles and the draft-tube are removed.

Abrasion fracture contributes more than 99.5 % to the total amount of generated fragments; most of the fragments are comparable in size with the original crystals, since less than 1 wt. % of the crystals reduce their size of 60 % or more.

The experimental CSDs can be satisfactorily fitted by a simple model combining abrasion and shatter fracture; the parameters of the model have a physical meaning and their values allow to evaluate the fracture behaviour of solid particles suspended in a liquid.

ACKNOWLEDGEMENT

The author is indebted to A. De Santis for her contribution to the experimental work. This work has been financially supported by a grant of MURST 60 % (Italy).

NOTATION

c = suspension density [kg/m^3]

DT+B = group formed by the draft-tube and the baffles

k_s = fraction of the overall fracture due to shatter mechanism

L = crystal size [mm]

L_{max} = maximum size of the crystals [mm]

n = stirring rate [1/s]

n_s = modulus of size distribution equation of shatter fragments

n_c = modulus of size distribution equation of coarse abrasion fragments

n_f = modulus of size distribution equation of fine abrasion fragments

x = L/L_{max}, dimensionless crystal size

x_c = dimensionless size of the smallest coarse abrasion fragment

x_f = dimensionless size of the largest fine abrasion fragment

y_w = cumulative wt. fraction of fragments smaller than size L

$y_{w,a}$ = cumulative wt. fraction of abrasion fragments smaller than x

$y_{w,s}$ = cumulative wt. fraction of shatter fragments smaller than x

y_{wf} = overall cumulative wt. fraction of fine fragments smaller than x_f

W = cumulative wt. percent

LITERATURE CITED

1. Nienow, A.W., *Trans. Instn Chem. Engrs.*, (london) **54**, 205 (1976).

2. Nienow, A.W. and R.Conti, *Chem. Eng. Sci.*, (London) **33**, 1077 (1978).

3. Kelly, E.G. and D.J. Spottiswood, Introduction to Mineral Processing, p. 113, John Wiley & Sons, New York, 1982.

4. Mazzarotta, B., *Chem. Eng. Sci.*, (London) **47**, 3105 (1992).

5. Budz, J., A.G. Jones and J.W. Mullin, *AIChE Symp. Ser.* **83**, 253 (1987).

6. Broadbent, S.R. and T.G. Callcott, *J. Inst. Fuel*, (London) **29**, 524 (1956).

7. Ottens, E.P.K. and E.J. de Jong, *Kristall Technik*, (Berlin) **9**, 873 (1974).

A Computational Study of the Flow Pattern in an Industrial Paper Pulp Chest with a Side Entering Impeller

André Bakker and Julian B. Fasano

Chemineer, Inc. 5870 Poe Ave., Dayton, OH 45414

A mathematical model for the combined laminar and turbulent flow of paper pulp has been developed. The results of the model predictions agree well with experimental data. The model is used to analyze the flow pattern in an industrial paper pulp chest equipped with a side entering impeller. Turbulent, laminar and stagnant regions can easily be located. The model is an excellent tool for the optimization of agitators for large industrial paper pulp chests.

Recently significant progress has been made in the application of Computational Fluid Dynamics (CFD) to the analysis of fluid flow in stirred chests. However, most work concentrated on the computation of flow patterns in lab-scale tanks equipped with top entering impellers.

Industrial problems are usually much harder to tackle. Complicating factors are the use of fluids with complex non-Newtonian behavior, the use of side entering agitators instead of top entering agitators and the fact that the agitator may be operated in the transitional regime.

A difficult mixing problem is found in the paper industries. The paper pulp, which is a suspension of thin, flexible fibers, exhibits a very complex rheology. As a result of this rheology multiple flow regimes are found in paper pulp storage tanks. Part of the chest will be laminar; some parts of the chest are turbulent.

Pulp storage chests can be rectangular or cylindrical. The bottom of the chest is usually filleted, sloped and/or curved. Although paper pulp chests are sometimes equipped with top entering agitators, the paper industry prefers using side entering agitators.

The curved bottom shape and the side entering agitator require the use of body fitted grids for CFD modelling of this type of stirred tanks. The fact that multiple flow regimes occur makes modelling difficult since there are no well tested models for this type of flow available.

This paper deals with the development of a mathematical model for the flow of paper pulp in stirred tanks. First the model will be described. The model predictions will then be compared with experimental results for a lab-scale tank. Finally, the model will be used to predict the flow pattern in an industrial chest.

RHEOLOGICAL PROPERTIES OF FIBER SUSPENSIONS

The rheological properties of fiber suspensions are discussed by Gullichsen (1985). Figure 1 shows a typical shear stress vs shear rate diagram. This diagram shows that the fiber suspension initially behaves as a non-newtonian fluid with a shear stress τ_y. When the shear stress exceeds a certain value τ_d the fiber network structure is disrupted and the suspension behaves nearly as a turbulent Newtonian fluid.

As a result of this rheological behavior fiber suspensions are extremely difficult to agitate. To provide motion through the whole tank the shear stress has to exceed the yield stress everywhere in the fluid. Since gradients in shear stresses can be expected, there will be regions in the fluid where the fiber network structure is disrupted and the flow will be turbulent. At the same time the flow may be laminar or even be stagnant in other parts of the chest.

This combination of turbulent flow and

laminar flow of a non-Newtonian fluid makes this a difficult to model problem with CFD. As far as the present authors know, nothing has been published about the direct computation of this type of flow in stirred tanks.

DESCRIPTION PAPER PULP MODEL

The computational model is based on solving the conservation equations which describe the flow in the tank. For the mean flow of an incompressible liquid, the equation for the conservation of mass reads:

$$\frac{\partial U_i}{\partial x_i} = 0 \tag{1}$$

U_i denotes the mean velocity in direction x_i. Under steady-state laminar flow conditions the momentum balance is given by:

$$\rho U_j \frac{\partial U_i}{\partial x_j} = -\frac{\partial p}{\partial x_i} + \frac{\partial \tau_{ij}}{\partial x_j} + \rho g_i \tag{2}$$

The stress tensor τ_{ij} is given by:

$$\tau_{ij} = \mu\left(\frac{\partial U_i}{\partial x_j} + \frac{\partial U_j}{\partial x_i}\right) \tag{3}$$

When the flow is turbulent an additional term appears in the momentum balance for the mean flow:

$$\rho U_j \frac{\partial U_i}{\partial x_j} = -\frac{\partial p}{\partial x_i} + \frac{\partial}{\partial x_j}(\tau_{ij} + \rho \overline{u_i u_j}) + \rho g_i \tag{4}$$

u_i is the fluctuating, turbulent, velocity component in direction x_i. The cross correlation term $\overline{u_i u_j}$ denotes the Reynolds stress tensor. Although analytical equations for the Reynolds stress tensor can be derived, these equations again contain higher order cross correlation terms, and the set of equations is not closed. For modeling the Reynolds stresses various so-called turbulence models are available. An often used model is the k-ϵ model which models the Reynolds stresses as being proportional to gradients in the mean flow.

$$\rho \overline{u_i u_j} = -\mu_t\left(\frac{\partial U_i}{\partial x_j} + \frac{\partial U_j}{\partial x_i}\right) + \frac{2}{3}\delta_{ij} k \tag{5}$$

The turbulent viscosity μ_t is given by:

$$\mu_t = C_\mu \rho \frac{k^2}{\epsilon} \tag{6}$$

The model equations for the conservation of the turbulent kinetic energy density k and the turbulent energy dissipation rate density ϵ are:

$$U_k \frac{\partial k}{\partial x_k} = \frac{\partial}{\partial x_k}\left([\nu + \nu_t/\sigma_k]\frac{\partial k}{\partial x_k}\right) + P_k - \epsilon \tag{7}$$

$$U_k \frac{\partial \epsilon}{\partial x_k} = \frac{\partial}{\partial x_k}\left([\nu + \nu_t/\sigma_\epsilon]\frac{\partial \epsilon}{\partial x_k}\right) + C_{\epsilon 1}\frac{\epsilon}{k}P_k - \epsilon$$

Equation (5) for the Reynolds stresses is similar to equation (3) for the laminar shear stresses. The turbulent viscosity μ_t plays the same role as the molecular viscosity μ. As a result the form of the turbulent momentum equations is similar to the form of the laminar momentum equation except that μ is replaced by an effective viscosity μ_{eff}:

$$\mu_{eff} = \mu + \mu_t \tag{9}$$

For modelling the fiber suspension, which is partly laminar and partly turbulent, the following method was used. For every computational cell the computations are first performed as if the flow is turbulent. After that it is checked if the total shear stress is indeed larger than τ_d

$$(\mu + \mu_t) S \geq \tau_d \tag{10}$$

The shear rate S is calculated as the square root of the scalar invariant of the shear rate tensor:

$$S = \sqrt{\frac{\partial U_i}{\partial x_j}\left(\frac{\partial U_i}{\partial x_j} + \frac{\partial U_j}{\partial x_i} - \frac{2}{3}\delta_{ij}\frac{\partial U_l}{\partial x_l}\right)} \tag{11}$$

If equation (10) is not satisfied

$$(\mu + \mu_t) S < \tau_d \tag{12}$$

then the calculations for that particular cell are repeated as if the flow is laminar. The local apparent viscosity is then calculated from the experimental shear rate - shear stress curves and the local shear rate S.

The CFD solver used in this work is Fluent V4.11. Fluent contains the k-ϵ model for turbulent flow, but does not contain any models for combined laminar and turbulent flow. The additional equations were coded in Fortran subroutines and linked to the Fluent solver. The solver was then configured such that these additional subroutines were called every iteration.

EXPERIMENTAL VERIFICATION

Usually the results of CFD models are verified by performing flow visualizations, eg. with the aid of Laser Doppler velocity measurements or other techniques, and comparing the results with the model predictions. Here this is not possible since it is not possible to see through the fiber suspension.

To allow for easier flow visualization a model fluid was designed. The model fluid consists of a suspension of glass fibers in a mixture of Tetrachloroethylene (TCE) and Carbondisulfide (CS_2). The concentrations of TCE and CS_2 were chosen such that the refractive index of the mixture matched the refractive index of the suspended glass fibers. The resulting mixture was clear enough to be able to see through and to spot the motion of tracer particles, despite a 10% volume holdup of glass fibers.

The experimental set up consisted of a cylindrical glass vessel with a diameter of 0.3 m equipped with a 0.12 m Chemineer HE-3 impeller, operating at a speed of N = 10.8 1/s (= 650 RPM). The experiments were performed with a top entering impeller. The fact that the model will later be applied to tanks with side entering agitators is not expected to be a problem.

White tracer particles were added to the tank. A video film of the experiment was made and pictures were taken. The results of the flow visualization experiments were then compared with the results of the model calculations.

Figure 2 shows the model predictions for this tank. The model predicts three zones in the tank: turbulent flow, laminar flow and stagnant. The region around the impeller is characterized by high velocities and is turbulent. This is the only region where the shear stresses are large enough to disrupt the fiber network structure. Around this turbulent zone there is a zone where the fiber suspension is moving but where the stresses are not large enough to disrupt the fiber network and where the flow will be creeping laminar. Even further away from the impeller, in the corners at the bottom and in the top of the tank the suspension is not moving at all.

These predictions compare well with the experimental results. The experimental results are summarized in Figure 3. In the experiments it was not possible to distinguish between laminar and turbulent flow. The experiments show larger stagnant zones than the model predictions do. However, in most regions where the flow should be stagnant the model predicts very low velocities. The overall comparison between the model predictions and the experiments is satisfactory.

INDUSTRIAL CHESTS

A typical industrial paper pulp chest is shown in Figure 4. The sides of the chest are 10 m each and the liquid level is 7 m. The bottom of the chest is curved. The side entering impeller is mounted near the draining point. This is the generally accepted procedure in the paper industries.

The agitator studied here was a 1.22 m (= 48") diameter Chemineer HE-3S impeller running at 4.67 1/s (= 280 RPM). The computer model was used to study the flow pattern generated by this agitation system for various pulp consistencies. The consistency is defined as the mass fraction of bone dry paper fibers.

The shear stress vs. shear rate curves used here were the curves published by Gullichsen for Softwood Kraft pulp. Other types of paper pulp have will have different rheological behavior depending on, amongst other things, fiber length and diameter.

The body fitted computational grid (see Figure 4) was generated with Fluent PreBFC V4.01 and then exported to Fluent V4.11 for the computations and the post processing.

The tank was divided in 30x30x21 = 18,900 internal cells. Average computation time per simulation was 12 hours on an HP 9000-750 workstation.

RESULTS

Figure 5 shows the predicted velocity vectors in a plane through the impeller for turbulent flow of water. The jet coming from the impeller impinges on the opposite chest wall and a long circulation loop is formed.

Figures 6 and 7 show the velocity vectors in the same plane for pulp consistencies from 1% and 3.5%. The size of the circulation loop decreases with increasing pulp consistency.

The length of the arrows is proportional to the local velocity. Short arrows denote low velocities; long arrows denote large velocities. No arrows at all means that the velocities are too small to be displayed, although they are not necessarily zero.

Figure 8 shows a wireframe surrounding the volume around the impeller where the shear stresses are large enough to disrupt the fiber network structure and where the flow will be turbulent (5% consistency). Figure 9 shows a similar wireframe, but here the wireframe encloses the volume where the velocities are larger than 1E-3 m/s. The region outside the wireframe can be regarded as stagnant.

Figure 10 shows a velocity distribution diagram. The volume fraction of fluid moving with a velocity larger than a certain minimum velocity is plotted for various consistencies. This diagram shows that this agitator provides motion control for consistencies which are lower than 5%, which agrees well with the results of the Chemineer design procedures. The velocities of 1E-3 m/s may seem low at first, but creeping laminar flow is usually all that is needed in this type of storage chests.

CONCLUSIONS

For the first time a model has been developed for the multi-regime flow of a fluid with a complex rheology like paper pulp. The results of the model calculations agree satisfactory with experimental flow visualizations.

The model can be used to predict the flow pattern in large industrial chests where zones with turbulent mixing, zones with laminar mixing and stagnant zones can easily be located. This model is an excellent tool for the optimization of agitators for large industrial chests.

LITERATURE

Gullichsen J. (1985)
Medium Consistency Processing - Fundamentals
Bleach Plant Operations/TAPPI Seminar Notes, 135-142

SYMBOLS

$c_{\epsilon 1}$	Model Constant
$c_{\epsilon 2}$	Model Constant
g	Acceleration by Gravity
k	Turbulent Kinetic Energy Density
N	Impeller Rotational Speed
p	Pressure
P_k	Production of k
S	Shear Rate
u	Fluctuating Component of Velocity
U	Mean Velocity
x	Coordinate
ρ	Density of Fluid
δ	Dirac Delta function
σ_ϵ	Model Constant
σ_k	Model Constant
τ	Shear Stress
τ_d	Disruptive Shear Stress
τ_y	Yield Stress
μ	Molecular Viscosity
μ_t	Turbulent Viscosity
μ_{eff}	Effective Viscosity
ν	μ/ρ
ν_t	μ_t/ρ

Figure 1 Typical Shear Stress Shear Rate Diagram

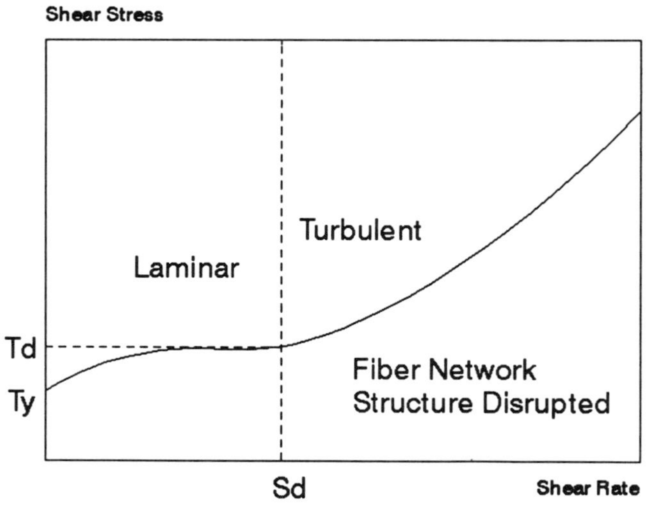

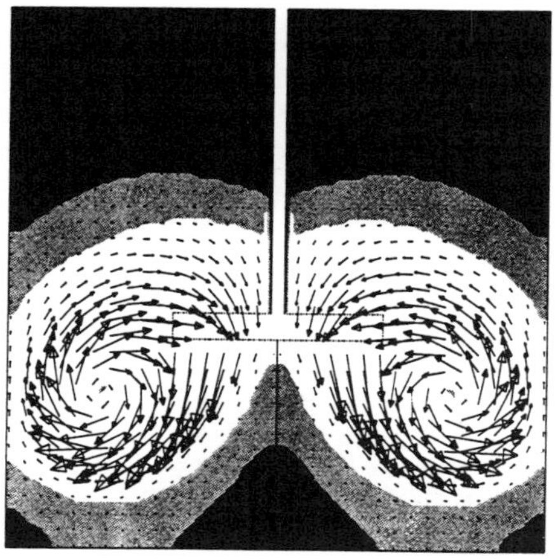

FIGURE 2 - Model Predictions.
Black: Stagnant; Grey: Laminar;
White: Turbulent;

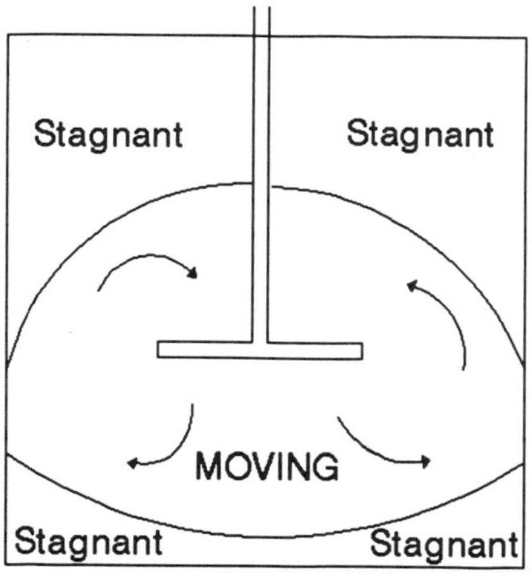

FIGURE 3 - Experimental Visualization

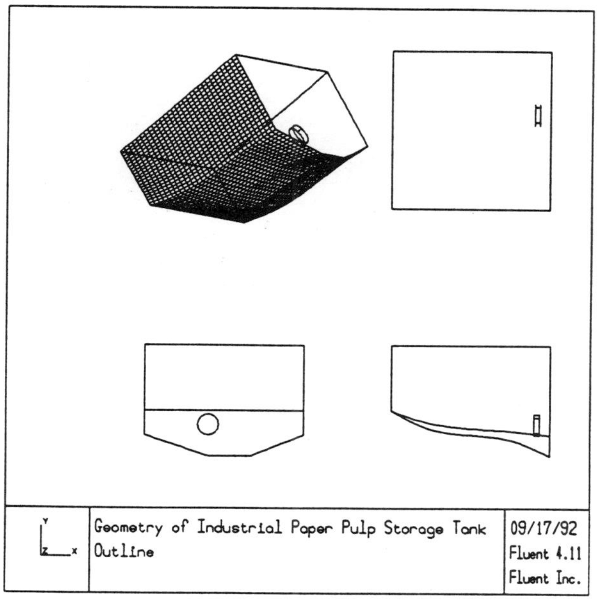

Figure 4 - Outline and Grid

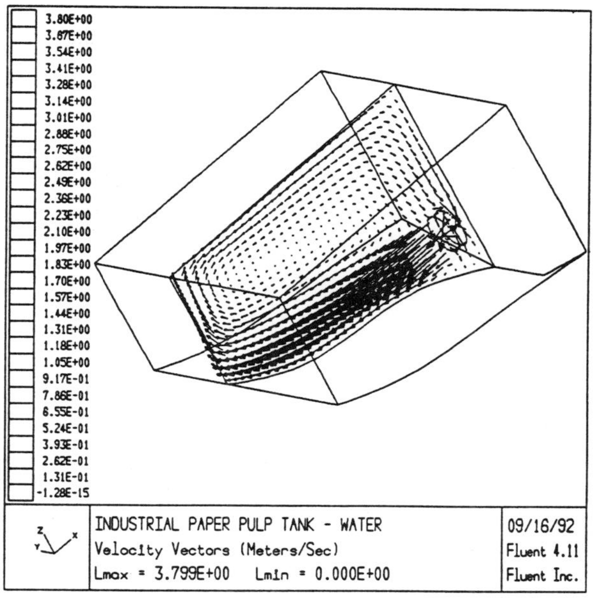

Figure 5 - Velocity Vectors (Water)

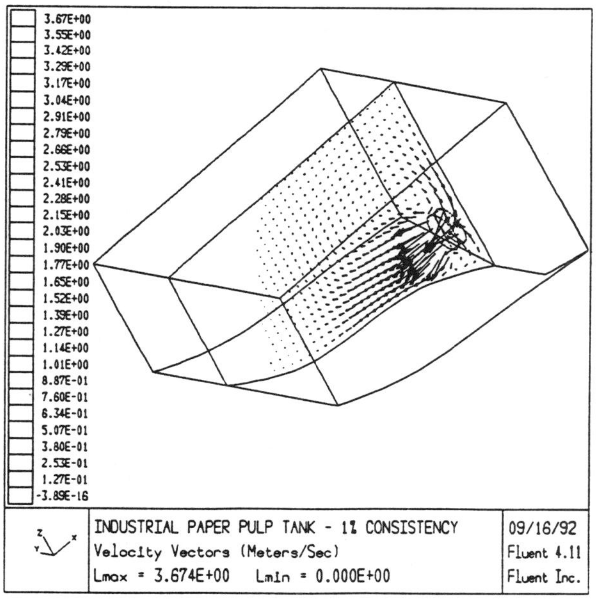

Figure 6 - Velocities, 1% Pulp

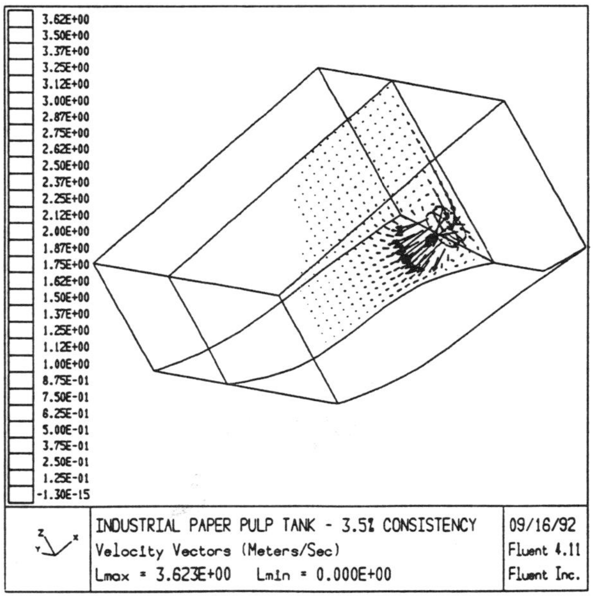

Figure 7 - Velocities, 3.5% Pulp

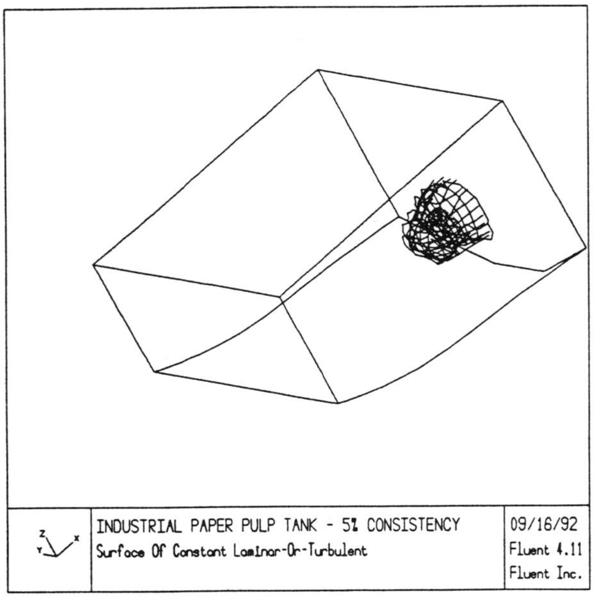

Figure 8 - Turbulent Volume (5% Pulp)

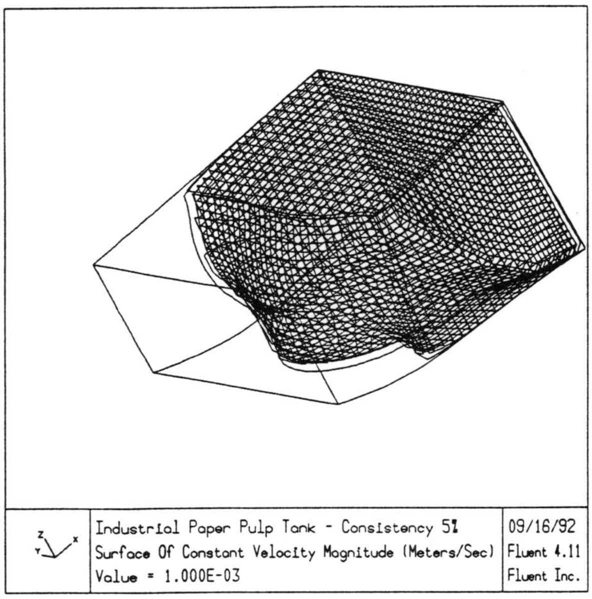

Figure 9 - Volume with velocities larger than 1E-3 m/s (5% Pulp)

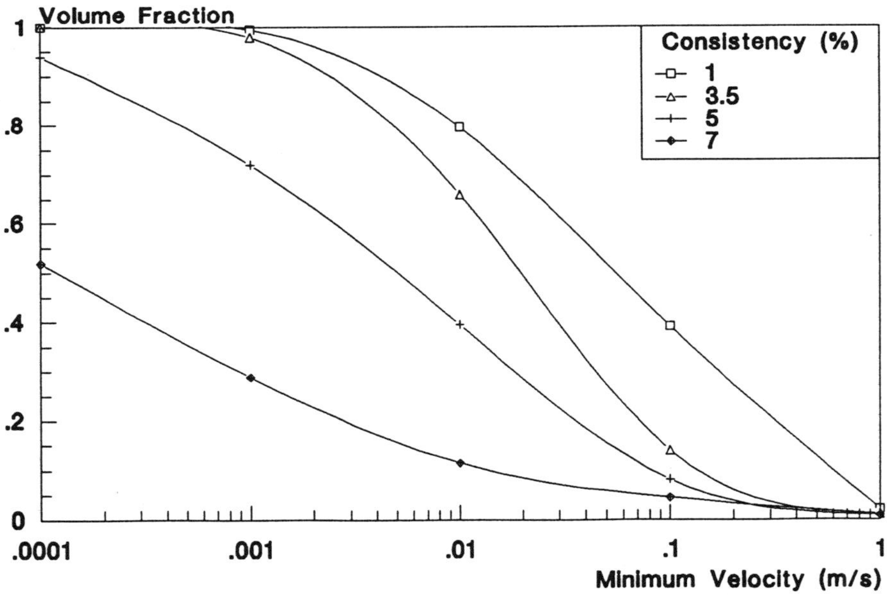

Figure 10 - Volume Fraction as a Function of Minimum Velocity (m/s)

Development of a Mathematical Model for Evaluation of Blending of Stratified Miscible Liquids

Kulin Lodaya, Russell J. Allgor, John W. Mitchell, and Anthony J. Zehnder

Air Products & Chemicals Inc. Allentown, PA 18195

A mathematical model has been developed which quantifies the blending times in low viscosity miscible liquid/liquid, flow sensitive systems. Liquid blending can take appreciable time in systems with significant density differences or strict requirements on the final degree of homogeneity. The total blending time is assumed to be the sum of: (1) the time to break any stratification caused by large density differences, using an algorithm dependent on the Richardson number, (2) the time to blend a pulse of the lightest liquid to 99% homogeneity using a correlation based on the Reynolds number, (3) the additional time required to achieve the desired degree of homogeneity assuming the deviations follow an exponential decay from 99%. Parameters for the three submodels were determined using data taken at an Air Products operating facility manufacturing proprietary surfactant blends, where blending time had become a bottleneck in the process. The model has been applied to a variety of systems throughout Air Products. Agreement with operational experience is quite good were data exists.

In the manufacturing of surfactant blends at an Air Products domestic manufacturing facility, the time required to blend the components of the final product before packaging had become a bottleneck. The standard practice for the mixing of these products was to blend each of the mixtures for four hours in a 6000 gallon (10.5 foot outside diameter, with four-bladed 45° 10 inch wide 42 inch diameter axial flow impeller, and four extended baffles) tank before packaging and shipping. We were certain that this amount of time was not required for all of the blends. However, premature removal of a blend results in off-spec material and attendant lost time to reblend the product. If more were understood about the factors governing the mixing process, the risk of removing a batch too soon could either be eliminated or greatly reduced.

In the initial work of analyzing the blending problem, we evaluated several correlations available in the literature for liquid blending by comparing the mixing time predicted by these equations with the required blending times of a few of our products that had been tested. We found that all of the tested correlations significantly under-predicted the blend times of some of our two-component blends.

We examined correlations based on tank turnovers, variance decay, and empirical formulas built on dimensionless numbers describing the agitation in the tank, and found that the inadequacies of these correlations emanate from two sources. The first source of error originates in the way the correlations account for property differences between the fluids. The second shortcoming was that most of the correlations tested did not include the final variance of the blend as a variable in determining the time required to reach the "well mixed" state. The conclusion we reached was that Air Products requires a much more homogeneous blend than is commonly considered "well mixed" by the standard correlations.

Air Products and Chemicals, Inc., Allentown, Pennsylvania (USA)

On the basis of these initial evaluations, we began to develop a model that could account for:

1. The high degree of homogeneity required for our blends.
2. The physical property differences between the fluids to be blended together.

DISCUSSION

Modeling Approach and Model Requirements

Density differences between the components in some of our blends are quite large. In fact, because of the way the components are added to our blend tank, many of the mixtures are stratified during the early stages of mixing. The blending of these liquids is essentially a flow-controlled impeller-driven mixing process with mostly low viscosity fluids. Although the viscosity of the mixture will have a significant effect on the time required to blend, viscosity differences between individual components should have little effect on the required time. Viscosity differences are not considered in our model.

The Air Products definition of homogeneity requires that every container (down to a one gallon grab sample) taken from the tank contains the same composition. This definition corresponds to a degree of homogeneity in excess of 99%. In order to account for the differences between the definition that we have established for "well mixed" and the definition that many of the available correlations use, we decided to use a variance decay model similar to the one proposed by Levenspiel [5].

To describe both the major effects of stratification and the strict tolerances that we employ on the degree of homogeneity, we chose to model the mixing of the fluid as if it consists of two distinct physical steps. The first of the two steps is the mixing required to break any stratification(s) existing in the tank. The second step determines the time required to blend a pulse of fluid into a relatively homogeneous liquid to the degree of homogeneity desired. The second step follows logically from the first in that the pulse of fluid that is introduced to the relatively homogeneous liquid is actually the final portion of the least dense fluid that was originally floating on the top.

The following sections detail each of the submodels employed. By approaching the problem in this stepwise fashion, the time required to complete each of the steps will be calculated. With this information, the reasons for the unexpectedly long blend times (compared to the commonly used correlations) can be attributed to either the extra time it takes to get to the degree of homogeneity that we require or to the time it takes to break the stratification(s).

Submodel Descriptions

Stratified Blending Model. One well-known correlation in the mixing literature which attempts to describe stratified blending is that developed by James Oldshue of Mixing Equipment Company [8]. Oldshue assumes that the blending time is a function of the buoyant forces. Experiments with hot and cold salt water solutions were used to determine the effect of density difference on blending time. He plotted the blend time versus the ratio of the density difference to the blended density. The data was fit to a line on a log-log plot with a slope of 0.9 to determine his density difference correction factor. The equation is given as

$$\text{Blend Time} = K * (D/T)^{-2.3} * (HP)^{x1} * (\Delta\rho/\rho)^{0.9}$$

where
x1 = -1 for stratified blends
x1 = -1/3 for non-stratified systems.

A limitation of this correlation is that the density

differences examined were only in the range of 0.3% to 7%. Additionally, the line appears to be a good fit only at the lower end of the density scale, with the correlation underestimating the blend time for a mixture with a difference of 7% by a factor of two. Since some of our more difficult surfactant blends have density differences on the order of 30-40%, it is not surprising that this correlation fails to accurately predict blend time.

Oldshue provides a qualitative description of the mechanism as "a gradual erosion or wearing away of the barrier by the fluid in the part of the tank moved by the mixer." Under these conditions, a blend was not obtained until "the interface reached the surface of the liquid." This mechanism is different from the standard mixing mechanism where the entire tank can be in a state of turbulence. Oldshue also states that "the wearing away of the interface seems to be largely proportional to the volume of fluid passing across the interface" implying that increasing the pumping capacity will help speed up the process.

The blending of stratified fluids has been widely investigated in the field of oceanography. An excellent description of the fundamental mechanisms is given in J. S. Turner's book, "Buoyancy Effects in Fluids" [15]. He suggests that the mixing "takes place largely through a process which looks like the intermittent breaking of steep forced internal waves, which tends to thicken the interface, followed by the sweeping away of this fluid by the stirring in the layers, which sharpens the interface again."

Many of the experiments relating to stratified blending (Rouse & Dodu [10], Hopfinger and Toly [4], Thompson and Turner [13], Reilly and Prandtl [9], Smith and Schoenmakev [11], Takahashi et al. [12], Van de Vusse [16], and Wessenlingh [17]) have used water and salt water solutions to create the stratified layers. The entrainment velocity in all of these experiments was correlated as a function of the overall Richardson number. The Richardson number is a ratio of the buoyant and turbulent forces (it can also be viewed as the ratio of the increase in potential energy due to mixing to the kinetic energy supplied by the agitator). It is defined as:

$$Ri_o = g\,\Delta\rho\, l_1 / \rho\, u_1^2$$

Correlating the results with the Richardson number has been done over the last 20 years, and the relative entrainment velocity is generally accepted as a function of the Richardson number raised to the (-3/2) power. The scales used to determine the Richardson number should be "the integral length scale of the turbulence l_1, and the r.m.s. value of the horizontal component of the turbulent velocity u_1, both measured at the interface" (Turner 1986 [14]). The work of Xuequan E and Hopfinger [18] bears out this correlation, and mentions the work of others that has led to the acceptance of the 3/2 power law.

Although the relation determining the entrainment is a simple power function, the method is difficult to apply to our conditions because the length and velocity scales are not easily computed with the complex turbulence generation pattern associated with an impeller. Experiments ([4], [6], [13]) have shown that the mixing is a function of the local turbulence, not a function of the convective patterns. If we assume that the length scale for the problem is proportional to the large scale eddies in the tank it can be taken as a constant. Since this constant will be fit to our experimental data we should have a reasonable approximation for the length scale built into our model.

Hopfinger and Toly [4] also calculated the r.m.s. turbulent

velocity as a function of the grid parameters. Since the velocity correlation developed is specific to the method of mixing, it cannot be used to calculate the turbulent velocities for our model, but a rough approximation of the horizontal component of the velocity can be derived using geometric arguments. Since the pumping capacity of the impeller is determined from its size and speed, the volumetric flowrate can be used to determine an average velocity by assuming an arbitrary flow path for the fluid.

The horizontal component of the turbulent interfacial velocity is estimated using the computational fluid dynamics code FLUENT [3]. Since the pumping number identifies the volume pumped by the impeller, knowing the cross-sectional area of the impeller, the average velocity exiting from the impeller can be determined. This is used as the boundary condition for the FLUENT runs, and then the velocity profiles throughout the tank can be calculated. These profiles are not exact, but we need only get an estimate of the horizontal velocity at the interface as the interface moves. FLUENT runs were made for tanks filled to different heights and then the integral average of the horizontal velocity was calculated. The average velocities at different heights were then correlated to yield a function expressing the average horizontal velocity at the interface as a function of height.

This velocity correlation relates the velocity at the interface as a function of the interface position and the average velocity of the fluid exiting from the impeller. This accounts for any changes in the speed of the impeller. However, changing the diameter affects the flow pattern in the tank significantly; to properly account for this, additional FLUENT runs are required to allow the average horizontal velocity to be expressed as a function of interface height and D/T ratio. Figures 1 and 2 illustrate the idealized flow patterns in our model of the stratified tank. As previously noted by Oldshue [8], the layer with the impeller is stirred and the stirring gradually erodes the interface.

We decided to model the blending rate as a function of the relative entrainment rates. The model assumes that a stratified fluid is created when the tank is filled. This should provide a conservative estimate of the blend time. The relative entrainment rate is calculated as a function of the Richardson number (unless the value of Ri_o is less than 7 which indicates that another mechanism may dominate):

$$E = u_e / u_1 = K\, Ri_o^{-3/2} = K (g \Delta\rho\, l_1 / \rho\, u_1^2)^{-3/2}$$

For the experiments performed by Xuequan E and Hopfinger [18], this form was used and the exponent determined as 1.54 +/- .05 at high Richardson and Peclet numbers. They also determined K to have a value of approximately 3.8. In our model the integral length scale of turbulence, l_1, is taken as constant for a given tank as discussed earlier. This constant is lumped with the K parameter to yield the relative entrainment rate as:

$$E = u_e / u_1 = K_1 (g \Delta\rho / \rho\, u_1^2)^{-3/2}$$

In our model, it was necessary to determine K_1 from data for each tank/impeller combination. This is accomplished by integrating the above equation over time, for a fixed value of K_1, until the interface reaches the top of the lightest layer. The value of K_1 is then varied until the integrated times match the experimental blend times.

Further insight into the stratified blending phenomena is given by Ahmad et al [1]. Similar to Oldshue, they described the mixing mechanism as "the turbulent erosion of the primary interface." The blending time was correlated against the Richardson number for density differences of up to 10%. However, they indicate that "at a sufficiently high rotation speed (they used an impeller rather than a grid), the primary interface is quickly broken by the liquid discharged from the propeller. The lighter liquid is mixed rapidly with the denser lower layer." This indicates that there is a limit to the applicability of the stratified blending mechanism. It makes sense that if enough turbulence could be generated to pull the upper layer into the impeller region that the mixing mechanism would change dramatically. In Turner's 1986 article [14], he indicates that this change in mechanism occurs for the range of $Ri_o < 7$.

Pulse Blending Model. As turbulence erodes the density interface, the interface moves to the top of the stirred fluid. During this movement we assume that the composition at any point in the stirred layer is the blended composition of the fluids originally in that volume. We use a pulse blending model to simulate the time it takes to blend a pulse of fluid into the system that is relatively homogeneous. This time is known to be a function of the power input to the tank, the pumping capacity of the impeller, the size of the system, and the degree of homogeneity.

We examined two existing correlations. The first is an equation developed by Oldshue [7], that predicts the blend time as a function of impeller power number N_p, D/T ratio, Reynolds number, and the density difference of the components.

$$\text{Pulse Blend Time} = K_2(N_p)^{x1}\,(D/T)^{x2}\,(\Delta\rho/\rho)^{0.9}\,(\Delta\mu/\mu)^{x3}$$

where

x1 = -1{for stratified}, -.33 {for non-stratified}

x2 = -1 to -2 {function of fluids, geometry...}

x3 = constant determined via experiment

$N_p = Pg_c/\rho N^3 D^5$

For our blends the viscosity exponent x3 was nearly zero, indicating that the viscosity difference correction was unimportant. The time predicted by Oldshue's correlation corresponds to a degree of homogeneity of 95%. This correlation predicts the blend time as a function of the power draw on the impeller, which is in turn dependent on the power number and the size and speed. The power number, typically expressed as a function of the Reynolds number, approaches an asymptotic value at Reynolds numbers greater than about 10,000.

The other correlation we considered was developed by Chemineer [2] and directly relates the blend time to the Reynolds number:

$$\text{Pulse Blend Time} = K * (1/N) * (D/T)^{-2.3} * f(Re).$$

One major difference between the Chemineer and Oldshue correlations is that the Reynolds numbers at which an increase in the Reynolds number has no effect on the blend time is smaller for the Oldshue correlation. The power number approaches an asymptotic value at a Reynolds number of about 10,000, but our data indicate that the blending time remains a function at Reynolds numbers greater than 10,000. The difference in the Reynolds number at which the power numbers and blending times approach asymptotic values occurs because the power number is a function of only the local turbulence at the impeller. The blending time is a function of the turbulence existing throughout the tank. The blending time function does not reach an asymptote until the entire tank is at a state of high turbulence. The simplified graphs taken from Chemineer [2] demonstrate that the mixing time and power numbers tend to an asymptote at different values of the Reynolds number.

For these reasons the pulse blending time in our model uses a

Chemineer type correlation. We chose a function that uses a constant that we can fit to the model to account for our definition of "well mixed" and for the fact that at the end of the stratification the batch may not be at exactly the initial conditions used to generate the curve:

$$\text{Pulse Blend Time} = K_2\ (1/N)\ (D/T)^{-2.3}\ f(Re)\ (\Delta\rho/\rho)^{0.9}$$

Variance Decay Model. The variance decay model adjusts the pulse blend time for varying degrees of homogeneity in the final blend. The model is based on Levenspiel's exponential decay model [5], and was adapted to be consistent with the pulse blending model. Figure 3 shows conceptually how the submodel operates. If h_p is the homogeneity calculated from the pulse blending at time t_p , we can apply a decay model to predict the homogeneity at times other than t_p as follows:

$$h = 1 - (1-h_p)e^{-K_3 t}$$

where
t = additional time required to make the specified blend
h = desired homogeneity (fraction)
K_3 = decay constant, calculated as follows:

$$K_3 = \ln[(1-h_s)/(1-h_p)]/t_p$$

In most of our work, the stratified homogeneity, h_s, was taken as 0.7, and the pulse blend homogeneity, h_p, was taken as 0.99.

RESULTS

Table 1 displays a comparison of model predictions to data collected for several of our blends. Overall, the agreement is quite good. Of particular interest is the fact that the model was able to accurately predict not only total blend time, but also accurately predicts the time to break the initial stratification. While more testing is warranted, it appears the model may be applicable to a wide variety of flow-sensitive blending processes.

A PC program has been developed which predicts the blending time using the formulations described in this paper. The PC program is written in FORTRAN and has been tested on machines with minimum capacity equivalent to an IBM 286 based machine.

CONCLUSIONS

A mathematical model was developed to predict the time required to produce a homogeneous blend of any number of miscible liquid components. The model accurately describes the process, even when large density differences cause initial stratification of the system. The algorithms employed within the model are general in nature and can be used to predict the blending times for any flow sensitive mixing system. This model can be used by both the plant operators, to see how long a particular charge to the blending tank should take to become "well mixed", and by the process engineers to evaluate how changes in the mixing equipment are likely to affect the blending times.

NOMENCLATURE

D = impeller diameter.

E = relative entrainment velocity.

g = gravitational constant.

h = desired homogeneity (fraction).

h_p = homogeneity from pulse blending model.

h_s = homogeneity from stratified blending model.

K_1 = correlation constant for stratified blending model.

K_2 = correlation constant for pulse blending model.

K_3 = correlation constant for variance decay model.

l_1 = the integral length scale of the turbulence.

N = agitator speed.

N_p = power number.

P = agitator power input.

Re = Reynolds number.

Ri_o = Richardson number.

T = tank diameter.

t = time in variance decay model.

t_p = pulse blending model time to homogeneity.

u_1 = r.m.s. value of the horizontal component of the turbulent velocity.

u_e = entrainment velocity.

x1 = power number exponent in pulse blending model.

x2 = geometric exponent in pulse blending model.

x3 = viscosity exponent in pulse blending model.

Greek Symbols:

$\Delta\rho$ = density difference between stratified layers.

ρ = blended fluid density.

$\Delta\mu$ = viscosity difference between stratified layers.

μ = blended fluid viscosity.

LITERATURE CITED

1. Ahmad, S. W., B. Latto, and M. H. I. Baird, "Mixing of Stratified Liquids", Chemical Engineering Research and Design, vol. 63, pp. 157-167, 1985.

2. CHEMINEER, "CE Refresher: Liquid Agitation", selected reprints from Chemical Engineering (1975-76).

3. FLUENT User's Manual, TN-369, Rev. 1., Creare Inc.

4. Hopfinger, E. J., and Toly, J. A., "Spatially Decaying Turbulence and its Relation to Mixing Across Density Interfaces", J. Fluid Mech., 60, 467, (1976).

5. Levenspiel, Octave and Soon J. Khang, "The mixing-rate number for agitator-stirred tanks", Chemical Eng., October 11, 1976, pp. 141-143 (Oct. 1976).

6. Mory, M., and Hopfinger, E. J., "Rotating Turbulence Evolving Freely from an Initial Quasi-2D State",from Macroscopic Modelling of Turbulent Flows, Lecture Notes in Physics, Vol. 230, Springer (1985).

7. Oldshue, J. Y., AIChE Meeting, Philadelphia, 1989.

8. Oldshue, J. Y., Hirshland, and Gretton, "Blending of low-viscosity liquids with side entering mixers", Chem. Eng. Prog., vol. 52, 11, pp. 481-484. (Nov 1956).

9. Rielly, C. D., and A. B. Prandtl, Proc. 6th Eur. Conf. on Mixing, Pavia, Italy, BHRA Fluid Eng., Cranfield, England, vol 69, 1988.

10. Rouse, H., and Dodu, J., "Turbulent Diffusion Across a Density Discontinuity", Houille Blanche, 10, 522, (1955).

11. Smith, J. M., and A. W. Schoenmakev, Chem. Eng. Res. Des., vol 66, p. 16, 1988.

12. Takahashi, K., Y. Takahata, T. Yokota, and H. Konno, J. Chem. Eng. Japan, Vol. 18, p. 159, 1985.

13. Thompson, S. M. and J. S. Turner, "Mixing across an interface due to turbulence generated by an oscillating grid", J. Fluid Mechanics, vol. 67, pp. 349-368 (1975).

14. Turner, J. S., "Turbulent Entrainment: the development of the entrainment assumption, and its application to geophysical flows", J. Fluid Mechanics, vol. 173, 431, 1986.

15. Turner, J. S., Buoyancy Effects in Fluids, London: Cambridge University Press, 1973.

16. Van de Vusse, J. G., Chem. Eng. Sci., vol. 4, p 178, 1955.

17. Wessenlingh, J. A., Chem. Eng. Sci., vol. 30, p. 973, 1975.

18. Xuequan E and E. J. Hopfinger, "On mixing across an interface in stably stratified fluid", J. Fluid Mechanics, vol. 166, pp. 227-244 (1986).

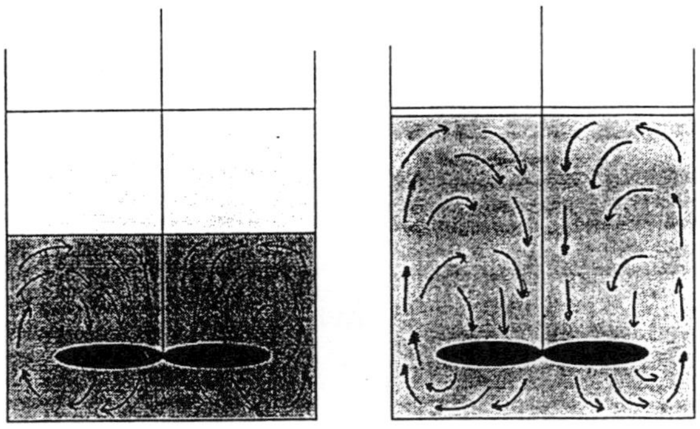

Figure 1. **Stratification Model**

1. Assume blend is stratified at time zero
2. Calculate the movement of the density interface

Table 1.

Comparison of Model to Operating Plant Data

	Experimental Time, minutes		Model Prediction, minutes	
Blend	Stratification	Total	Stratification	Total
A	15	45	15	46.5
B	5	45	6	43.5
C	?(1)	<15(2)	1.5	10.5
D	<5(1)	120	4	120

(1) Experimental time to break stratification was too short to measure.
(2) Total blend time was estimated.

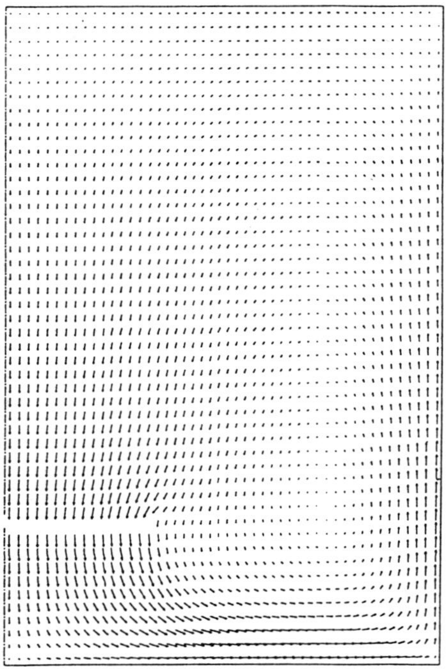

Figure 2. Velocity Profiles calculated from FLUENT

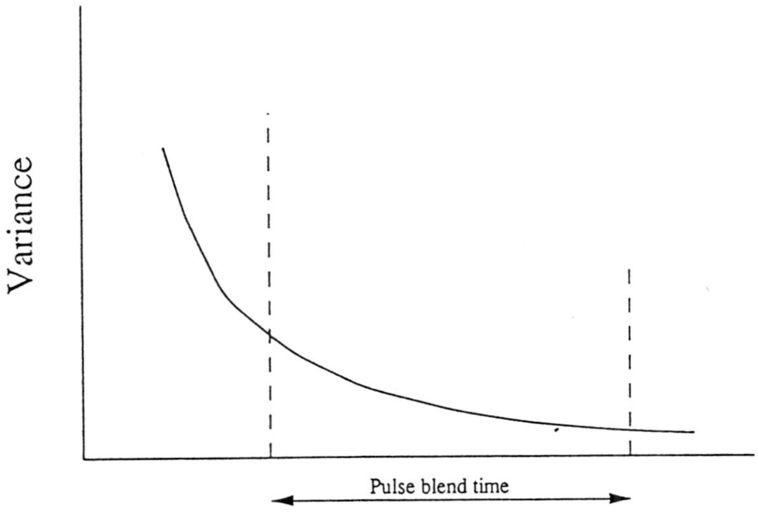

Figure 3. Variance Decay Model

Determine exponential decay constant from the pulse blending time

Calculate time for any degree of homogeneity

Cocontinuity and Percolation in Correlated Structures

E. Bruce Nauman

The Isermann Dept. of Chem. Eng., Rensselaer Polytechnic Inst., Troy, NY 12180

Percolation theory is extended to mixtures with nearest neighbor interactions. A pair interchange model simulates second order correlations in two and three dimensions. Phase diagrams show extended regions of cocontinuity for the normal case of agglomerating phases. Simulations indicate that systems containing three to five cocontinuous phases can be physically achieved using more-or-less random mixing or phase separation processes suited to large scale manufacturing.

INTRODUCTION

Percolation theory studies the connectivity of a component in a random mixture. As a simple example, consider a large grid of squares with individual squares being colored black or white according to some probability rule. If the proportion of white in the overall grid is high, most of the white squares will have one or more white sites as neighbors, forming clusters of the white phase. If the white phase represents voids and the black phase represents an impenetrable solid, then a fluid can percolate through a arbitrarily large grid if the fraction of white sites exceeds about 59%. An alternative interpretation is that the white phase is continuous when the phase fraction - the area fraction in this two dimensional example - exceeds about 59%.

Percolation theory was introduced by Broadbent and Hammerlsey (1957). Two aspects of a percolation problem are the dimensionality and the shape (and corresponding coordination number) of the space filling element. The above example was in two dimensions and used square elements with a coordination number of 4. The element for percolation in one dimension is a point, and the critical percolation threshold is 100% since a single impenetrable point will block percolation on a line. Critical percolation thresholds in two ore more dimensions depend on the shape of the space filling elements. The present study treats square elements in two dimensions and cubic elements in three dimensions with coordination numbers of four and six, respectively.

Analytical solutions are notoriously difficult to obtain in percolation theory. The percolation threshold for triangular elements in two dimensions is known to be 50% (Sykes and Essam, 1964). Simulation has been used for most problems, but accuracy is difficult to achieve by this method. For squares in two dimensions the time sequence of estimates of the percolation threshold is 55% (Domb and Sykes, 1961), 58.1% (Frisch, et al., 1961), 58% (Dean, 1963), 59.0% (Sykes and Essam, 1964), 59.1% (Dean and Bird, 1967), 59.275% (Stauffer, 1985), and 59.3% (Huff, 1991). Estimates for cubes in three dimensions by the same authors follow a similar time series ending at 31.4% (Huff, 1991). Huff has reviewed various methods for estimating percolation thresholds, the more sophisticated of which are based on moments of the cluster size distribution. Note that an cluster of infinite size is formed at the percolation threshold.

Interest in percolation theory has peaked and waned. It was an exciting topic in the physics literature in the 1960's. A survey article (Mohanty, Ottino and Davis, 1982) and a text (Nauman and Buffham, 1983) introduced percolation concepts to the chemical engineering literature, and a monograph (Stauffer, 1985) has appeared. It is an interesting mathematical concept, but physical applications of classic or Bernoullion percolation theory are limited.

The present paper aims an rekindling interest in the materials engineering aspects of percolation theory and multiphase mixtures. Plastics having two continuous phases offer an unusual balance of properties compared to the more usual situation of one continuous phase and one dispersed phase. In what range of phase fractions can two continuous phases be produced by mixing operations or by phase separation processes? Can material systems with three or more cocontinuous phases be produced by bulk methods? Conventional percolation theory gives rather unexciting answers to these questions. The percolation threshold of 31.4% for cubes in three-space suggests that cocontinuous mixtures are feasible for phase fractions from about 1/3 to about 2/3. A three-phase, cocontinuous mixture is theoretically possible but just barely so. Note that our question specifically involves cocontinuity of two or more phases and not just the lower limit on continuity of a single phase. Single phase continuity can be observed down to arbitrarily low phase fractions, witness the water phase in packed, wet sand. See also Fizazi, et al. (1990). The water/sand example can also be used to envision cocontinuity at arbitrarily low phase fractions of the minor component since one can mentally imagine connections between the sand particles which do not destroy the continuity of the water phase. Such structures may be possible and have practical utility, but are unlikely to be formed by more-or-less random mixing or phase separation processes most suited to large scale manufacturing.

Classical percolation theory uses a particular probability rule for assigning phases to individual elements: each assignment represents an independent probability event. This is called Bernoullian percolation and implicitly defines non-Bernoullian percolation as having element assignment rules that utilize conditional rather than independent probabilities. Most real mixtures have structure. The structure may be caused by the molecular attractions in fluids. It may be caused by the mechanics of crack propagation in porous solids. The structure usually corresponds to a positive correlation where the existence of a phase at one site makes it more likely that the same phase will exist at an adjacent site, but negative correlations are also possible. Non-Bernoullian percolation theory models the correlated structures in real mixtures by using conditional assignment probabilities. It reduces to Bernoullian percolation for completely unstructured mixtures.

CONDITIONAL PROBABILITY ASSIGNMENTS

This section discusses methods for assigning phases to particular sites within a percolation space. We begin with Bernoullian percolation of a two phase mixture. Let P_A be the desired fraction of phase A. Generate a random sample from a uniformly distributed population having range 0 to 1. If the sampled random number is less than P_A, assign phase A to the site. Otherwise assign phase B to the site.

If the size of the space is large, the above procedure will give a realized phase fraction which is close to the desired value, P_A. Consider a square or cubic space containing L^d, d = 2 or 3, sites. Then let N_A equal the integer nearest $P_A L^d$. The realized phase fraction can be made exactly equal to N_A/L^d by adding or deleting phase A from a relatively few of the initially selected sites. The site locations for these additions or deletions are chosen at random.

The above procedure generates an uncorrelated structure. For large L^d, pick pairs of adjacent sites at random. Let P_{AA} denote the fraction of the selected pairs where both sites contain phase A and so on. Then

$$P_{AA} = P_A^2 \qquad P_{AB} = P_{BA} = P_A(1-P_A) \qquad P_{BB} = (1 - P_A)^2 \tag{1}$$

The above, of course, are theoretical expressions for uncorrelated structures. They can be closely approximated in a well designed simulation experiment. The probability P_A can be chosen subject only to the constraint that

$$1 > P_A > 0 \tag{2}$$

Applying this probability independently at each site automatically generates the above results for P_{AA}, P_{AB}, and P_{BB}. Indeed, these expressions can be regarded as the definition of an uncorrelated structure.

For structures with nearest neighbor correlation, P_A can again be chosen subject only to Equation 2. The probability of a pair, P_{AA}, can also be chosen independently but subject to the constraints

$$P_A > P_{AA} > 0 \tag{3a}$$

and

$$P_A > P_{AA} > 2P_A - 1 \tag{3b}$$

The pair probability for uncorrelated structures, $P_{AA} = P_A^2$, lies within these limits. If $P_{AA} > P_A^2$, the system shows positive correlation and tends towards large clusters of the A phase. If $P_{AA} < P_A^2$, the system shows negative correlation so that A sites tend to be isolated. If $P_A < 0.5$, the lower limit on P_{AA} is zero. At this lower limit, every A site will be isolated. If $P_A > 0.5$, the fraction of phase A is so large that some AA pairs must exist. The minimum number of such pairs corresponds to $P_{AA} = 2P_A - 1$. The lower limit on P_{AA}, whether 0 or $2P_A - 1$, can be achieved or closely approximated on a grid of finite size. The upper limit of $P_{AA} = P_A$ is theoretical possible for any P_A and corresponds to very large domains of the individual phases. This limit is difficult to realistically approximate with a finite grid.

Of the probabilities P_{AA}, $P_{AB} = P_{BA}$, and P_{BB}, only one is independent. Assuming P_{AA} is specified, the others are calculated from

$$P_{AB} = P_A - P_{AA}$$
$$P_{BB} = 1 - 2P_A + P_{AA} \tag{4}$$

If P_{AB} is specified, then

$$P_{AA} = P_A - P_{AB}$$
$$P_{BB} = 1 - P_A - P_{AB} \tag{5}$$

The constraints on P_{AB} are

$$\min (P_A, 1-P_A) > P_{AB} > 0 \tag{6}$$

We turn now to the problem of generating a percolation space which satisfies the nearest neighbor correlations indicated above. Specifically, we wish to specify P_A and P_{AA} and then generate a grid which achieves these probabilities to reasonable accuracy. This is easily done in one dimension (it is a Markov process) but is difficult in higher dimensions. No algorithm has yet been found that will directly generate conditional probability spaces without introducing anisotropy. Instead, we use an indirect method which is conceptually similar to the Ising model of theoretical physics. The procedure begins by generating a space with $P_A = 0.5$, $P_{AA} = 0$. This corresponds to a checker board pattern in two dimensions or a staggered set of checker boards in three dimensions. If the desired P_A is less than 0.5, sites originally assigned to phase A are reassigned to phase B until the desired P_A is achieved. Since $P_{AA} = 0$ initially, the random deletion of phase A sites maintains this value. If the desired P_A exceeds 0.5, phase A sites are added at random locations. At least for fairly large grids, this procedure gives a realized P_{AA} which closely approximates the minimum possible value, $P_A = 2P_A - 1$ for $P_A > 0.5$. See Figure 1.

To achieve P_{AA} higher than the starting value, pairs of sites are selected at random. If interchanging the contents of the sites increases P_{AA}, the interchange is made. Otherwise, it is not. Note that the sites are randomly chosen from throughout the entire grid while the Ising model limits interchanges to adjacent sites. The pair selection procedure continues until the desired P_{AA} is achieved. The procedure causes P_{AA} to increase monotonically, but the rate of increase becomes very slow as P_{AA} approaches its upper limit of P_A.

Figure 2 shows square grids with L = 40 and P_A = 0.5 for various values of P_{AA} as generated by the pair interchange algorithm. Figure 2a with P_{AA} = 0.1256 retains some of the checkerboard pattern of the initial configuration. This largely disappears in Figure 2b with P_{AA} = 0.2500 which corresponds to the Bernoullian case. The cluster size distribution has a somewhat smaller mean and a larger variance when the pair interchange algorithm is used to achieve $P_{AA} = P_A^2$ compared to direct Bernoullian generation. See Table 1. No significant differences in percolation thresholds were observed for these methods for achiving $P_{AA} = P_A^2$. Note that the pair interchange algorithm can be implemented starting from a Bernoullian base with $P_{AA} \approx P_A^2$ rather than from the checkerboard pattern with P_{AA} = 0.

CONTINUITY AND PHASE DIAGRAMS

Phase A is continuous if percolation on it is possible in all principal directions. In three dimensions, percolation paths must extend from each face of the cube to the opposite face. These paths need not belong to the same cluster. In two dimensions, percolation paths must extend from each side of the square to the opposite side. These paths must necessarily belong to the same cluster due to the nature of two-space.

The current study used the above, direct definition of continuity to estimate the phase diagrams shown in Figures 3 and 4. The results displayed in these figures were not computed to high precision but are believed accurate to about 0.01 probability units.

Figure 3 shows the two dimensional case. The primary independent variable is P_A, the phase fraction. The secondary independent variable is P_{AB}, which acts as a form of dispersion parameter. P_{AB} was used rather than P_{AA} because it gives symmetrical phase diagrams. When P_{AB} is above its Bernoullian value, P_A $(1-P_A)$, the phases tend to disperse rather than agglomerate, and the region in which neither phase is continuous is extended.

Table 1

Comparison of Nominally Uncorrelated Structures, $P_{AA} = P_A^2$, Generated by the Direct Bernoulli Method and by the Pair Interchange Model, P_A = 0.25, Cubic Lattice, L = 40, Average of 100 Trials.

	Fraction of Total Clusters	
Cluster Size	**Bernoulli**	**Pair Interchange**
1	.5911	.6680
2	.1428	.0928
3	.0717	.0625
4	.0431	.0394
5	.0292	.0256
6	.0211	.0183
7	.0157	.0139
8	.0120	.0108
9	.0097	.0087
10	.0080	.0071
Average Cluster Size	3.1607	2.9659
Standard Deviation	6.3145	7.7618
Average Maximum Size	138.3	203.5

In two dimensions, a highly dispersive system can have no continuous phase up to a phase fraction of about 71%. In an agglomerating system with P_{AB} below the Bernoullian value, continuity is enhanced. However, a phase cannot be continuous below a phase fraction of 50%. Cocontinuity is impossible in two dimensions when using the percolation definition of continuity as given above. (Other definitions are possible.)

Figure 4 displays the phase diagram for three dimensional percolation. There are four possibilities in a two phase system: Phase A continuous, phase B continuous, neither phases continuous, both phase continuous. Previous studies recognized only three possibilities. In Bernoullian percolation, phase B is continuous when the volume fraction of phase A is below 31%. Both phases are continuous when the volume fraction of phase A lies between 31 and 69%, and only phase A is continuous at volume fractions above 69%. Non-Bernoullian percolation offers the fourth possibility of neither phase being continuous. Figure 4 exhibits a region between 43 and 57 volume percent where neither phases is continuous in dispersive systems with high P_{AB} and low P_{AA}. The more interesting aspect of Figure 4 is the extension of cocontinuity to volume fractions less than 20% in agglomerating systems with low P_{AB} and high P_{AA}. This opens the theoretical possibility of creating materials having five or more continuous phases.

The boundary between single phase continuity and cocontinuous is shown as a dotted line for phase volumes above 80% or below 20%. It is not yet clear whether the ambiguity in this region represents an inherent aspect of the second order, conditional probability model or merely reflects a simulation difficulty. For phase fractions above 20% (but below 80%), the percolation transition is a sharp function of P_{AA} at a given P_A. Below 20% volume fraction, the transition is diffuse and the probability of percolation on a large ($L = 40$, $L^3 = 64{,}000$) but finite grid shows an interior maximum with respect to P_{AA}. See Figure 5. The physical explanation of this phenomenon seems fairly straightforward. High values of P_{AA} can be achieved with the extended and tenuous phase structures that give rise to cocontinuity or with compact, dispersed structures. At volume fractions below about 20%, the pair interchange models can generate either of these structures when the grid is finite. On an infinite grid, the structure generated by the pair interchange model is presumably deterministic for a given phase volume, being one of the two structures. It is tempting to extrapolate the (tenuous) phase boundary down to volume fractions less than 10%, but an earlier transition to a dispersed phase could occur.

MULTIPHASE CONTINUITY

Consider a system consisting of N_p phases: A, B, C, ... A conditional probability model that treats nearest neighbor interactions will involve the set of second order probabilities P_{AA}, P_{AB}, P_{AC}, ... , P_{BB}, P_{BA}, P_{BC}, ... These may be displayed in matrix form as

$$\mathbf{S} = \begin{pmatrix} P_{AA} & P_{AB} & P_{AC} & \cdots \\ P_{BA} & P_{BB} & P_{BC} & \\ P_{CA} & P_{CB} & C_{CC} & \\ & \vdots & & \end{pmatrix}$$

$$= \begin{pmatrix} P_{A/A}P_A & P_{A/B}P_B & P_{A/C}P_C \cdots \\ P_{B/A}P_A & P_{B/B}P_B & P_{B/C}P_C \cdots \\ P_{C/A}P_A & P_{C/B}P_B & P_{C/C}P_C \cdots \\ \vdots & & \end{pmatrix} \tag{7}$$

where $P_{A/A}$ is the conditional probability of finding the second member of a randomly chosen pair to be an A given that the first member of was observed to be an A, with similar definitions for $P_{A/B}$ and so on. Note that **S** is a state matrix which describes the distribution of nearest neighbors chosen at random. It is not the transition matrix that appears in the theory of Markov processes. The latter would be the matrix of conditional probabilities $P_{A/A}$, $P_{A/B}$, ... Isotropic, correlated spaces cannot be achieved by a Markov process when $d > 1$.

The matrices in Equation 7 are symmetric for the isotropic spaces under consideration. Thus $P_{AB} = P_{BA}$ and so on. They are also subject to the constraint that

$$\sum_{rows} \mathbf{S} = \begin{pmatrix} P_A \\ P_B \\ P_C \\ \cdot \\ \cdot \\ \cdot \end{pmatrix} \quad (8)$$

$$\sum_{columns} \sum_{rows} \mathbf{S} = 1 \quad (9)$$

This results in N_p - 1 independent, first order probabilities and N_p (N_p - 1)/2 independent, second order probabilities. Characterization of the mean composition and the various spatial correlations requires all these parameters to be known. There are many possibilities for $N_p > 4$, and some combinations can be expected to enhance or particularly to hinder cocontinuity. In a more specific physical model, the P_A, P_B, ..., might be related to molecular abundances and the P_{AA}, P_{AB}, ... to intermolecular attractions.

For present purposes, we assume the simplest possible case

$$P_A = P_B = ... = P_1 = \frac{1}{N_p}$$
$$P_{AA} = P_{BB} = P_{CC} = ... = P_{11} \quad (10)$$
$$P_{AB} = P_{BC} = P_{AC} = ... = (P_1 - P_{11})/(N_p - 1)$$

so that there is one adjustable parameter, P_{11}. This situation can be simulated using a variant of the pair interchange process discussed previously. If the interchange increases the total number of like neighbors, the exchange is made. The process causes a monotone increase in

$$P_{xx} = N_p P_{11} \quad (11)$$

where P_{xx} is the probability that the elements in a pair chosen at random are equal.

This simplest possible case was simulated for N_p = 3, 4, 5 using a cubic grid, coordination number 6, with L = 40, L^d = 64,000. The criterion for continuity required each of the phases to percolate simultaneously in all three principal directions. It was readily satisfied for N_p = 3 and 4 at values of P_{11} consistent with Figure 4: P_A = 0.33, P_{AA} = 0.10 for N_p = 3; P_A = .25, P_{AA} = 0.09 for N_p = 4. The criterion was more reluctantly satisfied for N_p = 5, presumably due to the limitation on L but possibly representing an underlying physical limitation for random processes of phase separation.

CONCLUSIONS

A physical chemistry text from my youth - outdated even then - showed a photograph of seven, coexisting liquid phases, the known limit. Polymer chemistry indicates an unlimited number of coexisting liquid phases. The new question becomes: how many liquid (or solid) phases can be cocontinuous? This is not an equilibrium question since cocontinuous phases are not equilibrium structures, but they may exist long enough to form useful materials or at least to allow measurements.

Two cocontinuous phases (interpenetrating networks) have become items of commerce in polymers. There is no apparent bound on the number of phases that can be cocontinuous just as there is no lower bound on the volume fraction of a single continuous phase. There may be a practical limit when the phases are formed by quasi-random mixing and phase separation processes. We suggest this limit exceeds two and set it as a challenge to the experimentalist to produce structures having three or more cocontinuous phases.

Spinodal decomposition is a phase separation mechanism that tends to generate cocontinuity. In the later stages of spinodal decomposition, the phase morphology enters the so-called scaling region. Here, the dimensionless shape of the phase boundaries becomes time invariant although the phase separation distance grows (algebraically) with time. Numerical experiments suggest that cocontinuity is possible for volume fractions below 30%. Thus three phase, cocontinuous systems may be directly achievable by spinodal decomposition. An even higher number of cocontinuous phases could result when (tenuous) phase agglomeration is enhanced by gentle stirring.

The theory and methodologies presented here can be expanded in many ways. The accuracy of Figures 3 and 4 could be improved using large L, moment methods, and grid size extrapolation as discussed by Huff (1991). The detailed structure created by the pair interchange model can be more closely compared to the Bernoullian percolation structure. It is postulated that the multiphase, correlated structures calculated from Equation 10 are the most likely to give cocontinuity, but this has not been proved. Perhaps most important, the fixed cubic lattice is unrealistic for fluids and amorphous solids. A Voronoi polygon model (Winterfield, Scriven and Davis, 1981), extended to three dimensions and with correlation should be more appropriate.

REFERENCES

Broadbent, S.R. and Hammersley, J.M., "Percolation Processes: I. Crystals and Mazes", *Proc. Camb. Phil. Soc.*, **53**, 629-641 (1957).

Dean, P., "A New Monte Carlo Method for Percolation Problems on a Lattice", *Proc. Camb. Phil. Soc.*, **59**, 397-410 (1963).

Domb, C. and Sykes, M.F., "Cluster Size in Random Mixtures and Percolation Processes", *Phys. Rev.*, **122**(1), 17-78 (1961).

Fizazi, A., Moulton, J., Pakbaz, K., Rughooputh, S.D.D.V., Smith, P., and Heeger, A.J., "Percolation of a Self-Assembled Network: Decoration of Polyethylene Gels with Conducting Polymer", *Phys. Rev. Letters*, **64**, 2180-2183 (1990).

Frisch, H.L., et al., "Critical Percolation Probabilities (Site Problem)", *Phys. Rev.*, **123**, 1566-1567 (1961).

Huff, C.A., "Monte Carlo Analysis of Critical Transitions on Percolation Mediums", M.S. Thesis, Rensselaer Polytechnic Institute, Troy, NY, 1991.

Monhanty, K.K., Ottino, J.M., and Davis, H.T., "Reaction and Transport in Disordered Composite Media: Introduction of Percolation Concepts", *Chem. Eng. Sci.*, **37**, 905-924 (1982).

Nauman,E.B. and Buffham, B.A., MIXING IN CONTINUOUS FLOW SYSTEMS, Wiley, New York, 1983.

Stauffer, D., INTRODUCTION TO PERCOLATION THEORY, Taylor and Francis, London, 1985.

Sykes, M.F. and Essam, J.W., "Exact Critical Percolation Probabilities for Site and Bond Problems in Two Dimensions", *J. Math. Phys.*, **5**, 1117-1127 (1964).

Webman, I., Jortner, J., and Cohen, M.H., *Phys. Rev. B*, **14**, 4737 (1976).

Winterfield, P.H., Scriven, L.E., and Davis, H.T., "Percolation and Conductivity of Random Two-Dimensional Composites", *J. Phys. C., Solid State Phys.*, **14**, 2361-2376 (1981).

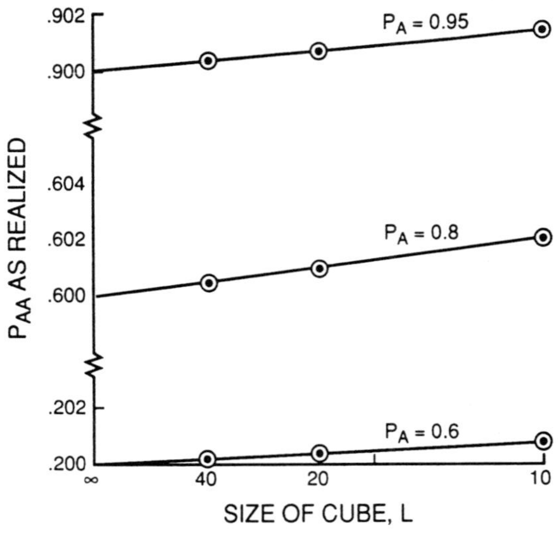

Figure 1. Convergence Toward Desired Correlation Probability.

a) $P_{AA} = 0.125$

b) $P_{AA} = 0.250$

c) $P_{AA} = 0.375$

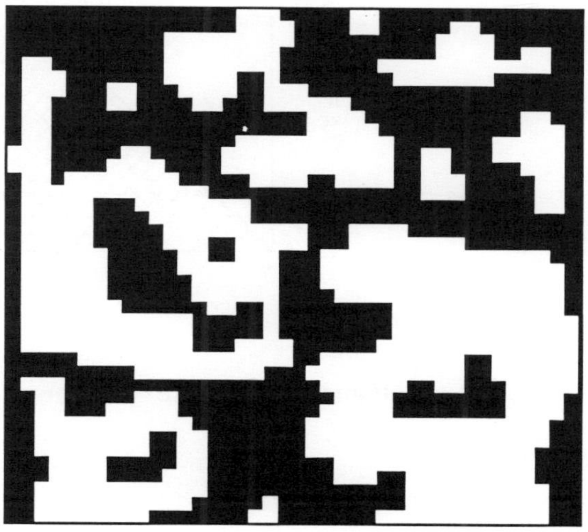

d) $P_{AA} = 0.425$

Figure 2. Random Structures with Correlation on a Square Grid with $P_A = 0.5$.

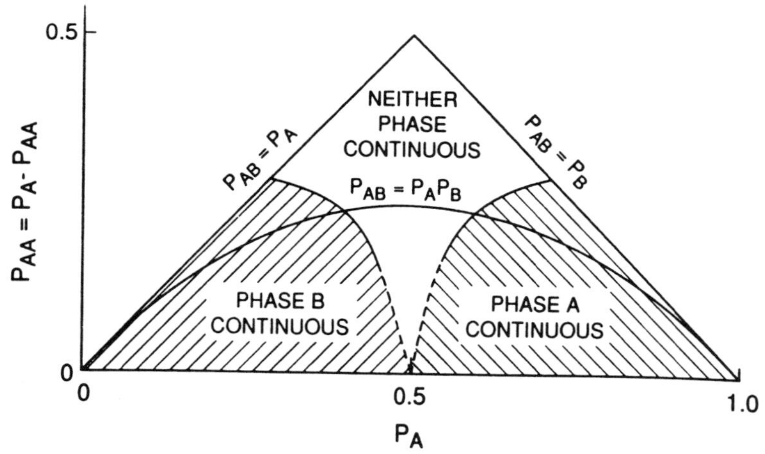

Figure 3. Phase Diagram for Two-Dimensional, Binary System.

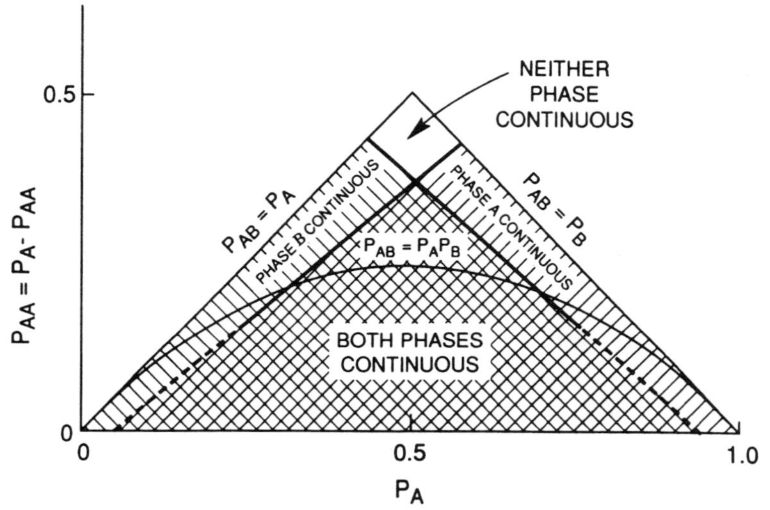

Figure 4. Phase Diagram for Three-Dimensional, Binary System.

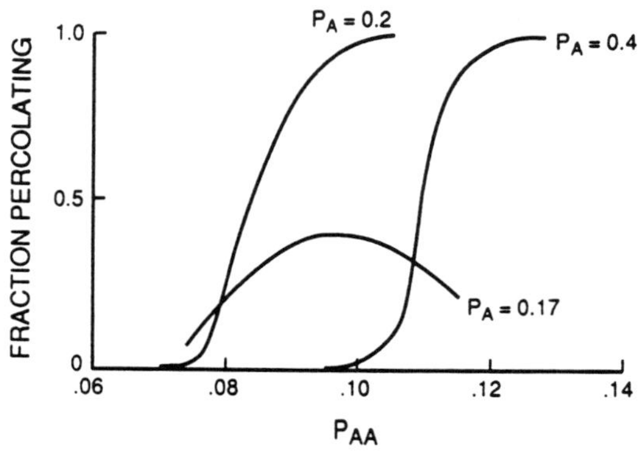

Figure 5. Percolation Fraction Versus Correlation Probability.

Dimensional Analysis, Similarity and Scale-Up

David S. Dickey

Prochem® Mixers, Robbins & Myers, Inc.,1895 West Jefferson St., Springfield, OH 45506

Dimensional analysis is the application of mathematical methods to physical problems. Desired results include fewer independent variables, evaluation of limiting cases, and size-independent correlations. Similarity provides a basis for creating and testing physical models, and justifies the methods for dimensional analysis. Scale-up is the design of processes or equipment based on small-scale testing. Rules and methods for scale-up may not follow directly from either dimensional analysis or similarity, but modeling often helps. No single procedure always works, only careful analysis will lead to successful scale-up.

The whole process of scale-up assumes that size-independent observations can be made over some range conditions. That range depends on the complexity of the process, the number of variables, the amount of previous experience and the accuracy required.

Dimensional analysis combines independent variables into logical parameters with mathematical or physical significance. Proper dimensional analysis should create size-independent relationships for process variables and reduce the number of independent variables to a minimum. The simplicity of dimensional analysis can also lead to serious errors if the variables are not selected or grouped correctly.

Similarity helps create models for physical systems, and establishes the range of conditions for testing. Similarity exists when corresponding ratios are the same in the model as in the full-scale. Similarity may result in equal values for dimensionless groups and provide size-independent results.

Scale-up is the process by which results obtained in a small-scale model are used to design a large-scale process or piece of equipment. Scale-up is usually intended to achieve a specific desired result.

* Robbins & Myers, Inc., Springfield, OH

The generalizations of both dimensional analysis and similarity may help determine how to adjust parameters for scale-up. Scale-up is not just similarity, and especially not equal dimensionless groups. Understanding the objectives of scale-up and limitations of dimensional analysis and similarity is essential.

BASIS FOR DIMENSIONAL ANALYSIS

Dimensional analysis begins with the premise that all relationships used to describe a physical system must have the same dimensions. For example, an expression can be written for the summation of forces at a point. However, the dimensions of each term in the expression must be force, or an equivalent combination, such as mass times acceleration. A valid expression cannot be written which combines length with time, or temperature with heat capacity, the dimensions must be the same for an equality to exist.

The basic dimensions of simple physical systems are length, time and mass. Other variables are combinations of units, i.e. velocity is length divided by time. Other dimensions, such as force, can be included in an analysis, but an additional variable must be included for each redundant dimension. The variable g_c has the dimensions mass times length divided by force and time squared, (ML/Ft^2).

Dimensionally consistent equations are directly related to physical similarity (similitude). The same concepts that justify physical models based on similarity also confirm dimensional analysis. However, dimensional analysis can be reduced to a mathematical theorem that defines the number of independent variables. The theorem is usually called the Buckingham Π-Theorem, since Buckingham [1] first proposed an organized approach to dimensional analysis based on similarity.

BUCKINGHAM Π-THEOREM
If there is a dimensionally homogeneous equation relating n quantities defined in terms of r reference dimensions, then the equation may be reduced to a relationship between $(n\text{-}r)$ independent dimensionless products Π_i, provided that the reference dimensions are independent.

References [2], [3], [4] and [5], each provide different definitions for Buckingham's Π-Theorem or a Π-Method. As a theorem, the number of dimensionless variables can be determined. As a method, appropriate independent groups are found. Whatever the definition or method, a properly posed analysis should lead to a minimum number of independent dimensionless groups.

The methods of dimensional analysis, are primarily mathematical and do not address the physical problem. The selection of appropriate variables, and the proper grouping of variables is not an assured outcome of the analysis.

DIMENSIONAL ANALYSIS

Dimensional analysis, when done correctly, is an extremely useful tool. The analysis should reduce the number of independent variables. Relationships between variables are established, and limiting cases can be evaluated.

The Buckingham Π-Theorem shows a reduction in the number of independent variables by at least the number of independent dimensions. A proper analysis will often combine the effects of physical properties. Fluid density and viscosity may appear as a ratio representing kinematic viscosity. Other variables, such as length dimensions, may appear as simple ratios. The length ratios resulting from dimensional analysis define geometric similarity. Thus, the small system is a scale model of the large one.

Limiting cases can be another very useful consequence of dimensional analysis. Typically, dimensionless groups express ratios relating different categories of similar phenomena, such as inertial and viscous forces. In a limiting case, such as very high viscosity, the viscous forces will dominate and the inertial forces become negligible.

Dimensional analysis can be conducted by several different procedures. Some procedures are very unstructured; others use some simple formulas or a systematic method of arranging variables. The analysis may begin with only the relevant physical variables, or it may begin with a mathematical model.

Any method of analysis is only as good as the starting point. Without a model, all of the relevant parameters must be correctly identified and appropriate groups formed. Omission of an important parameter can be disastrous. Inclusion of an inconsequential parameter often leads to improper conclusions and inaccurate correlations. Dimensional analysis with a mathematical model is more structured, but selection of an appropriate model is often difficult, especially the inclusion of boundary conditions for a complicated system.

ANALYSIS WITHOUT A MODEL

Dimensional analysis without a model begins with the selection of appropriate variables. Suppose an analysis were to be conducted on the thrust developed by a propeller, or mixing impeller. Thrust, $\mathcal{T}$ is the dependent variable and has the units of force, [F]. Reasonable independent variables for such an analysis include the following: axial velocity, v [L/t], impeller diameter, D [L], rotational speed, N [1/t], fluid density, ρ [M/L^3], fluid viscosity, μ [M/Lt]. The dimensions for the variables include length, [L], mass, [M] and time, [t]. The force dimension of thrust is a redundant dimension, formed from a combination of mass, length and time. A convenient method for handling force in the analysis is the introduction of another variable, g_c, which has the units of [ML/Ft^2].

The analysis begins with seven variables and four dimensions. Three independent groups can be defined based on the variables chosen for this analysis. Many dimensionless groups can be formed from the available variables:

$$\frac{T g_c}{\mu v D}, \frac{T g_c \rho}{\mu^2}, \frac{T g_c}{\rho N^2 D^4}, \frac{T g_c N^2}{\rho v^4},$$

$$\frac{T g_c}{\rho v^2 D^2}, \frac{T g_c}{\mu v^2}, \frac{T g_c}{\mu N D^2},$$

$$\frac{v}{ND}, \frac{D v \rho}{\mu}, \frac{N D^2 \rho}{\mu}, \frac{\rho v^2}{N \mu}$$

From these mathematical products several combinations of independent groups can be selected. One choice of groups would be

$$f\left(\frac{T g_c}{\rho N^2 D^4}, \frac{v}{ND}, \frac{N D^2 \rho}{\mu}\right) = 0$$

Other choices may be equally valid.

At this point, dimensional analysis reaches a critical limit. Further physical arguments about which groups are correct are only justification for experimental results. Experimental tests must be conducted to decide whether these are the appropriate dimensionless variables. Specific tests should be run to decide the most appropriate group involving thrust. Even good dimensional analysis must be verified experimentally.

Too many good experimental results are lost due to improper dimensional analysis. A perfectly good justification for a dimensionless group may not be confirmed by experimental results. Attempts to correlate experimental results with the wrong dimensionless groups often disguise the results and lead to erroneous conclusions about scale-up.

SERIES METHOD OF ANALYSIS

The series method **[3]** of dimensional analysis is another approach that does not use a specific mathematical model, but rather assumes that a general mathematical form will describe the process. This analysis is more systematic than the previous example, but still requires careful selection of appropriate variables and groups. The series method assumes that the dimensional relationship for a dependent variable, such as thrust, can be represented by an infinite series:

$$T = \alpha v^a D^b N^c \rho^d \mu^e g_c^f + \beta v^g D^h N^i \rho^j \mu^k g_c^l + \ldots$$

The coefficients, $\alpha, \beta, \ldots$, are dimensionless, so each term in the series must have groups of variables with the same dimensions as the result.

The exponents on individual factors must be selected to result in the units of force for the thrust expression. One of several combinations of variables provides the first term in the series:

$$\frac{T g_c}{v D \mu} = \alpha \left(\frac{DN}{v}\right)^c \left(\frac{D v \rho}{\mu}\right)^d + \ldots$$

Perhaps the most important aspect of this result is that this is only the **first** term in an infinite series. Often an analysis stops with this result, and correlation begins based on a log-log plot of the experimental results. A constant term, or a term with one exponent set to zero would be valid additional terms in this series.

Another error that is often made concerning the series method of dimensional analysis is plotting experimental results on log-log graphs. Because a large range of values can be depicted on such graphs, real variation can be disguised as if it were experimental scatter. Initial data should be plotted as a simple arithmetic function of the actual experimental variables. Then if the results show reasonable trends, the variables may be plotted as dimensionless groups.

ANALYSIS WITH A MODEL

Dimensional analysis, when conducted on a mathematical model, takes on the appearance of an analytical or mathematical operation. For instance, the equation of motion for a Newtonian fluid of constant density and viscosity is the Navier-Stokes equation **[6]**. In vector notation, the equation can be expressed as:

$$\rho \frac{Dv}{Dt} = -g_c \nabla p + \mu \nabla^2 v + \rho g$$

To simplify and generalize the solution of most any differential equation, the variables can be put in dimensionless form.

To analyze the Navier-Stokes equation, in terms of reference variables commonly used in mixing and agitation problems, the velocity is referenced to the product of rotational speed, N, and impeller diameter, D. A dimensionless expression for pressure can be formed using fluid density, ρ, along with speed and diameter:

$$p^* = \frac{(p - p_0)\, g_c}{\rho N^2 D^2}$$

With substitution of variables and extraction of coefficients, the Navier-Stokes equation can be put in the following dimensionless form:

$$\frac{\mathrm{D} v^*}{\mathrm{D} t^*} = -\nabla^* p^* + \left[\frac{\mu}{D^2 N \rho}\right] \nabla^{*2} v^* + \left[\frac{g}{D N^2}\right] \frac{\mathbf{g}}{g}$$

The coefficients are parameters of the equation, and they are also dimensionless groups. The coefficient on the velocity term is the inverse of the impeller Reynolds number, and the coefficient on the gravity term is the inverse of the Froude number.

An advantage of dimensional analysis with a mathematical model is that the method leads to specific groups. However, if the original equation is not entirely appropriate, or boundary conditions are not properly handled, the analysis is no better than simply picking variables.

Other dimensionless groups can be formed from the basic analysis. Suppose a relationship for impeller power, P, was sought. The resistance to impeller rotation, which is represented by power, can be assumed to be due primarily to the pressure differential across the blade. Thus, pressure acting on blade area, represented by D^2, times a moment arm, D, is torque. Torque times rotational speed, N, is power.

Substituting pressure represented as power into the expression for dimensionless pressure provides a new dimensionless group:

$$p^* \propto \left(\frac{P}{N D^3}\right) \frac{g_c}{\rho N^2 D^2} = \frac{P g_c}{\rho N^3 D^5}$$

This new group is the impeller power number, which is used to correlate power requirements for different impeller geometries.

Since the derivation of power number is from a dimensionless pressure distribution, the functionality developed by dimensional analysis should still hold.

$$\frac{P g_c}{\rho N^3 D^5} = f(N_{Re}, N_{Fr})$$

Again, dimensional analysis with a model provides insight into some assumptions and limitations necessary to derive a relationship.

LIMITING CASES

Dimensional analysis may also provide very useful information about what happens to functions as variables and groups approach limits. The previous analysis of the Navier-Stokes equation included a term for gravitational effects. While most mixing applications on earth are under the influence of gravity, there is rarely any significant difference in the magnitude or direction of the gravitational force. Thus if the force of gravity is always the same, and not a significant factor, it should be neglected.

A further simplification could have been developed if our original model had been Euler's Equation, which describes fluid motion, when both viscosity and gravity can be neglected. In dimensionless form, Euler's Equation reduces to the following expression:

$$\frac{D v^*}{D t^*} = -\nabla^* p^*$$

The absence of a parameter suggests that both the dimensionless velocity and pressure distributions are constants. If the dimensionless velocity is constant, the flow pattern and relative velocities are the same, even if the actual velocity changes.

If dimensionless pressure distribution is constant, then similar dimensionless groups are also constant. Thus, for low viscosity, the power number should be constant.

$$\frac{P g_c}{\rho N^3 D^5} = constant$$

Rearranging this expression shows that power is proportional to density, speed cubed and diameter to the fifth power:

$$P \propto \rho N^3 D^5$$

This result is observed experimentally.

PHYSICAL SIGNIFICANCE

An additional feature often ascribed to dimensionless groups is physical significance. Reynolds number involves parameters related to inertial forces in the numerator and parameters related to viscous forces in the denominator. Thus, Reynolds number effectively represents a ratio of inertial to viscous forces. The only real problem with this interpretation is that the magnitude of the forces is not truly represented, so a Reynolds number of one does not represent equal forces.

By interpreting dimensionless groups as ratios of physical characteristics, large and small magnitudes provide additional interpretation to experimental results. There are hundreds of other dimensionless groups that can be created for different problems, each with a potential significance to the physical problem **[7]**, **[8]**, **[9]**, and **[10]**.

SIMILARITY

The concept of similarity **[11]** was the topic that initiated the article by Buckingham **[1]** on dimensional analysis. Dimensional analysis and similarity have many common characteristics. Similarity defines the physical situation and dimensional analysis provides the mathematical justification. Both methods are aimed at creating or analyzing test results that can be more generally applied.

The first aspect of similarity is geometric similarity. In the most basic form, geometric similarity defines a scale model. A more general definition of geometric similarity is provided by Johnstone and Thring **[12]**.

GEOMETRIC SIMILARITY
For every point in one system there is a corresponding point in the other.

This definition of geometric similarity covers the traditional definition of geometric similarity when the scale ratios are the same in all directions. However, the definition also allows for distorted similarity, where geometric ratios may be the same in the x- and y-directions, but different in the z-direction. This definition also permits multiple corresponding points to exist. For instance a single bubble may be adequate to model a dilute dispersion of many bubbles. Any form of geometric similarity that is adequate to model essential aspects of a physical problem may be appropriate. However, geometric similarity with a single, scale ratio is easier to define and observe.

Geometric similarity is a prerequisite for other types of similarity. The geometric similarity must be adequate to model mechanical, thermal or chemical similarity for the chosen situation. A careful evaluation is necessary if distorted, geometric similarity is used.

Mechanical similarity is the next level of similarity above geometric similarity. Within mechanical similarity there are different forms.

MECHANICAL SIMILARITY

STATIC SIMILARITY
In static systems, corresponding stress results in corresponding deformation.

KINEMATIC SIMILARITY
In moving systems, corresponding particles move in geometrically similar paths in corresponding times.

DYNAMIC SIMILARITY
In moving systems, ratios of corresponding forces are equal.

Static similarity applies to solid objects where static conditions exist. Geometric similarity with a single ratio in all dimensions is usually necessary for static similarity. Kinematic and dynamic similarity apply to fluid systems. Fixed or distorted geometry may be adequate for similarity in some moving systems.

Another way of expressing kinematic similarity, is to say that the flow patterns are geometrically similar. When flow patterns are similar, heat and mass transfer rates should have a simple relationship in different sizes. Corresponding velocities and similar flow patterns are often essential features of fluid modeling. Dynamic similarity relates to forces acting upon and within a system.

THERMAL SIMILARITY
Corresponding temperature differences have a constant ratio, plus kinematic similarity for moving systems.

CHEMICAL SIMILARITY
With thermal similarity, corresponding concentration differences have a constant ratio, plus kinematic similarity for moving systems.

Thermal and chemical similarity are both more complicated than geometric and mechanical similarity, because the other similarities are prerequisites. Corresponding temperature differences with a constant ratio, mean that similar temperature distributions are achieved.

Chemical similarity requires geometric, kinematic and thermal similarity in addition to corresponding concentrations. Both thermal and chemical similarity are expressed in terms of differences, because temperature difference or concentration difference are the necessary driving forces, not absolute values.

True similarity with a single scale ratio is virtually impossible to use directly for scale-up. Some aspects of similarity provide a unique capability to correlate and generalize results. But the relative results and ratios of values are rarely the desired objectives of scale-up.

SCALE-UP

Scale-up usually requires that the magnitude of some variable or result be maintained. Scale-up is more a matter of selecting equal magnitudes than maintaining constant ratios. For scale-up it may be necessary to accurately duplicate agitation intensity in two different size vessels.

Many other scale-up criteria occur in mixing applications, including: equal impeller tip speed, equal torque per volume, equal power per volume, and equal solids suspension. Nearly all these criteria are easier to achieve with geometric similarity. The influence of unusual tank geometry on mixer performance can be very difficult to predict. Often a test of a scale model is the best way to investigate flow patterns.

Basic dimensional changes, such as impeller to tank diameter ratio and liquid level to tank diameter, can be essential geometry factors. Other variables, such as blade width or number of blades may be secondary, provided appropriate adjustments to rotational speed are made.

With geometrically similar mixers, and a single scale ratio, all large-scale dimensions can be based on the model. If a twelve-inch diameter tank is tested to duplicate a twelve-foot diameter tank, then similarly, but more important as a relative dimension, a four-inch impeller becomes a four-foot impeller.

Impeller diameter is more important than other dimensions because of the sensitivity of mixing problems to impeller diameter. In the turbulent range, impeller power is proportional to impeller diameter to the fifth power. Minor errors in both impeller diameter and rotational speed can greatly affect results. Other key dimensions must be within limited tolerances from exact ratios for scale-up formulas to accurately predict results.

If all of the dimensions are set by geometric similarity, the resultant variable for scale-up is agitator speed. Scale-up speeds can be predicted by a simple formula for relative speed.

$$N_2 = N_1 \left(\frac{D_1}{D_2} \right)^n$$

The scale-up exponent, n, determines the amount of adjustment that must be made to the agitator speed for different geometric ratios. The value of the exponent depends on the scale-up or design the criteria.

One common scale-up rule for agitation is equal liquid velocity. Equal velocity provides, what is commonly recognized as, equal agitation intensity for liquid blending problems. The scale-up exponent on the length-ratio factor for agitator speed is one.

$$N_2 = N_1 \left(\frac{D_1}{D_2}\right)^1$$

This scale-up rule is also called equal tip speed, because rotational speed is adjusted in inverse proportion to the length ratio. Because tip speed is a velocity, by similarity and dimensional analysis, it is related to the fluid velocities.

Torque per volume is another criterion used for some scale-up problems. Power is proportional to speed cubed and diameter to the fifth power. Volume is proportional to length cubed and impeller diameter cubed is an appropriate expression with geometrically similar systems. Torque is power divided by speed, so the relationship for power has the speed exponent reduced by one and the diameter exponent reduced by three.

$$\frac{\tau}{V} \propto \rho N^2 D^2$$

Because the exponents on both speed and diameter are the same, the scale-up exponent is again one. Torque per volume is equivalent to equal tip speed for geometrically similar systems.

Another scale-up method is equal power per volume. Equal power per volume may be appropriate for interfacial mass transfer conditions. Similar to torque per volume, power per volume can be developed for turbulent conditions as follows:

$$\frac{P}{V} \propto \rho N^3 D^2$$

Equal conditions in different sizes can be expressed as a ratio of the conditions in one scale compared to the conditions in the other.

$$N_2 = N_1 \left(\frac{D_1}{D_2}\right)^{\frac{2}{3}}$$

The scale-up exponent of two-thirds provides equal power per volume at turbulent conditions. Other scale-up conditions, such as equal blend time, result in other exponents.

A criterion with less definable conditions, such as equal solids suspension, may have an empirical exponent.

$$N_2 = N_1 \left(\frac{D_1}{D_2}\right)^m$$

The value of the solids suspension exponent is usually between two-thirds and one, with most typical values around 0.75 to 0.85. It is possible that different exponents apply for different types of suspension or different particles. A different exponent would suggest a different suspension mechanism, which could occur.

Most scale-up methods use some combination of dimensional analysis, similarity, magnitude equality, and engineering judgment. Different process applications also require different methods of scale-up **[13]**. Unfortunately, simple rules are less likely to apply to more complicated the systems. More complicated systems require a greater reliance on engineering judgment and experimental verification.

CONCLUSION

Successful scale-up takes advantage of dimensional analysis, and similarity. Dimensional analysis is just a tool, and is only as good as the user of that tool. Similarity sets limits within which modeling is possible. Scale-up is neither dimensional analysis nor similarity, but rather equivalent results.

Dimensional analysis identifies the appropriate parameters, but does not identify the form of the correlation. An infinite series with dimensionless groups as factors is one form. Similarity is an idyllic state in which variables can be independently adjusted to fit equal values for different dimensionless groups. Aspects of similarity are the starting point for both dimensional analysis and scale-up.

Scale-up is the result sought through dimensional analysis and similarity. The analysis and modeling form the basis for interpreting experimental results. Correlations and generalizations provide design relationships for larger systems. As with most engineering methods knowing both the possibilities and limitations is necessary to obtain successful results.

NOMENCLATURE

D	impeller diameter [L]
g	gravity acceleration [L/t^2]
g_c	gravitational constant [ML/Ft^2].
N	rotational speed [1/t]
N_{Fr}	Froude number
N_{Re}	Reynolds number
p	pressure [F/L^2]
P	power [FL/t]
t	time [t]
v	velocity [L/t]
V	volume
μ	viscosity [M/Lt]
Π	dimensionless product
ρ	density [M/L^3]
τ	torque [FL]
T	thrust [F]

Dimensions

[F]	force
[L]	length
[M]	mass
[t]	time

LITERATURE CITED

1. Buckingham, E., "On Physically Similar Systems: Illustrations of the Use of Dimensional Equations," *Phys. Rev.*, **IV**(4), 345 (1914).

2. de St Q. Isaacson, E. and M. de St Q. Isaacson, *Dimensional Methods in Engineering and Physics*, Halsted Press, John Wiley & Sons, New York, (1975).

3. Churchill, S. W., *The Interpretation and Use of Rate Data: The Rate Concept*, p. 202, Scripta Publishing Co., Washington, D.C., (1974).

4. Perry, R. H., D. W. Green, and J. O. Malony (Ed), *Perry's Chemical Engineers' Handbook*, 6th ed., pp. 2-113, McGraw-Hill, New York, (1984).

5. Baumeister, T., E. A. Avallone and T. Baumeister, III (Ed), *Marks' Standard Handbook for Mechanical Engineers*, 8th ed., p. 3-46, McGraw-Hill, New York, (1978).

6. Bird, R. B., W. E. Stewart, and E. N. Lightfoot, *Transport Phenomena*, p. 107, John Wiley & Sons, New York, (1960).

7. Boucher, D. F., and G. E. Alves, "Dimensionless Numbers," *Chem. Eng. Prog.*, **55**, 9, 55 (1959).

8. Boucher, D. F., and G. E. Alves, "Dimensionless Numbers - 2," *Chem. Eng. Prog.*, **59**, 8, 75 (1963).

9. Catchpole, J. P., and G. Fulford, "Dimensionless Groups," *Ind. Eng. Chem.*, 58, 3, 46 (1966).

10. Fulford, G. D., and J. P. Catchpole, "Dimensionless Groups," *Ind. Eng. Chem.*, 60, 3, 71 (1968).

11. Tolman, R. C., "The Principle of Similitude," *Phys. Rev.*, III(4), 244 (1914).

12. Johnstone, R. E., and M. W. Thring, *Pilot Plants, models and Scale-up Methods in Chemical Engineering*, McGraw-Hill, New York, (1957).

13. Bissio, A., and R. L. Kabel (Ed), *Scaleup of Chemical Processes*, John Wiley & Sons, New York, (1985).

Scale-Down Reactor Model to Simulate Insufficient Mixing Conditions During Fed Batch Operation Using a Biological Test System

Gen Larsson, Stefan George, and S-O. Enfors

Dept. of Biochemistry and Biotechnology, Royal Inst. of Tech., S-100 44 Stockholm, Sweden

A two-compartment reactor set-up, composed of a STR and a PFR, between which medium and cells are circulated, has been constructed. In this construction, the reactors can by choice be either aerated or not. By location of the feeding point, either in front of the plug flow part or in the stirred part, different types of gradients can be created in the system. The use of this reactor set-up is shown to influence the microbial metabolism, corresponding to substrate gradients in a large reactor. The Baker's yeast process was chosen as a model system and studied aerobically up to a biomass concentration of 38 g/L. By location of the feed point to the inlet of the scale-down reactor, a high concentration sugar region was created in the plug flow unit. This resulted in a six percent lower cell mass concentration and a higher ethanol production than in a control caused by the overflow metabolism of the yeast cells. This shows the importance of the location of the feeding point in a large scale reactor on the process outcome.

Calculation of the bioreactor performance will be of importance during scale up. While this situation is faced only when a new process is designed, scale down, i.e. the situation when new strains or raw material are tested in a small scale, is a far more frequent situation to be taken into account. A third occasion, when the bioreactor performance has to be considered, is in the design of new bioreactors. The strategy in obtaining scientific information for use in these situations can be of different kind. There has, for example, been suggested that scale up could be performed by a regime analysis i.e. by scale down of processes, based on critical parameters and their appropriate time constants. These would be tested and characterised in a small scale and on this characterisation the scale up would be based (Sweere et al, 1987).

In the Nordic Research Programme on Bioprocess Engineering, of which this work is a part, the bioreactor performance research is divided in three. First, a hydrodynamic part describs the mixing and mass transfer behaviour of the gas-liquid system of any reactor configuration. Second, the kinetics of the microbial behaviour in large reactors are needed. As a third part, the integration of these is considered in order to get the overall performance. In this paper, mainly the second part is described.

Substrate concentration gradients are likely to appear during large scale fed batch fermentation as a cause of insufficient mass transfer, insufficient mixing and a high microbial activity. Furthermore, these gradients can be of a different kind in different parts of the reactor. If the concentration of glucose for example is high in one area, because of feeding of substrate solution to the reactor in this part, then another limitation, in this case of oxygen, might occur. The existence of these gradients will create a constantly shifting environment for the cells. These gradients may or may not affect the microbial metabolism. The questions are: Will this alter the microbial reactions and what is the effect of this possible change on parameters like productivity, yield and product quality? The result of a cultivation under such circumstances, can in the best case only be the sum of the performance of the different parts of the reactor. It is very likely however that microbial reactions occur which will harm the process. How can this be investigated and the kinetics of this reactions be determined ?

It is the aim of this paper, to describe the strategy in finding the microbial response to fluctuating conditions and also to discuss briefly different reaction systems that could be affected under such conditions. As an example of this strategy, the Baker's yeast fed batch process is chosen and the short term effects of high sugar concentration on the cellular metabolism will be studied.

CONCENTRATION GRADIENTS

The gradients in a large reactor will, during a fed batch cultivation, be of different character. While one aims at a concentration, uniform over the reactor, in the range of some tenth's of milligrams per litre, in other parts, hundreds of grams per litre of feed solution is pumped into the broth. Under these conditions, gradients in the range of these values could theoretically be created. It is therefore possible, that the microbial environment will oscillate between excess and starvation and anything in between. Furthermore, the time scale of these fluctuations will vary when the cells makes a pathway through the reactor. How large these fluctuations appear, depends on the reactor geometry, the operating technique and conditions, but also on the microbial system in question.

	AREA A	AREA B	AREA C
CONC. IN CIRCULATION LOOP	"HIGH"	"LOW"	"AIMED"
CONC. GRADIENT CAUSED BY:	INSUFFICIENT MIXING	INSUFFICIENT MIXING HIGH CONSUMPTION RATE	"IDEAL" CONDITIONS
UNKNOWN CONSUMPTION RATE PROFILE:	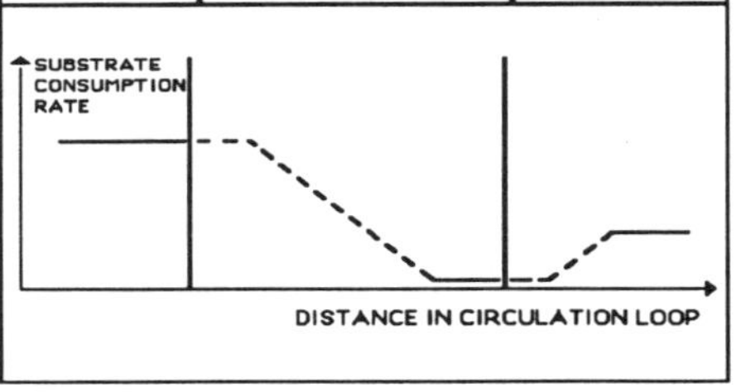		

Fig 1

A microbial reaction towards concentration gradients is schematically shown in fig. 1 above. The figure symbolises three different concentration areas and a possible effect on the substrate consumption rate. Under the special circumstances of an oscillating environment, the cellular response depends on the conditions last experienced. The response time and magnitude will therefore depend on this status of the microbial metabolism i.e. depending on earlier experience, on the concentration, the residence time in the area of interest and the microorganism in question.

THE SCALE DOWN COUPLED REACTOR CONCEPT

By design of scale down equipment, consisting of a series of reactors of different types, i.e. stirred or plug flow reactors, an experimental set-up can be created. This is done with the purpose of either to resemble a simple model of the large reactor or to give well defined conditions for a microbial kinetic evaluation. A suggestion of different reactor types are given below.

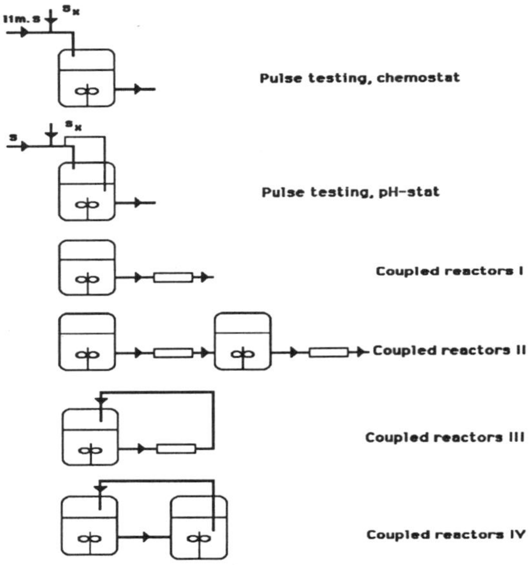

Fig 2

The reactor set-ups at the top of fig 2, symbolises continuous cultivations, but with different control schemes. The pH-auxostat works at the maximum growth rate, while the traditional chemostat works with a limitation of the growth rate by some nutrient component. In this type of equipment a number of metabolic studies have been performed, both of shift-ups and shift-downs in substrate concentration. These shifts are either of steady state or of oscillating character. If a change in the medium composition is made in a steady state operated chemostat, a number of residence times has to pass before a new steady state is reached. During this time, microbial changes of other character than the studied can occur, e.g. due to a genetic instability. This is one of the limitations with this operation. Others are, problem of wall growth in the reactor, cell culture synchronization and low feed rate caused substrate oscillation during steady state conditions.

During nonsteady state operation, the concentration varies with time. It is therefore not easy to achieve the microbial kinetics towards a defined residence time in an area of a certain concentration. However, the situation of a continuously shifting environment is probably a valid simulation of the nature of reactor gradients.

The four reactor configurations shown above in figure 2 (the coupled reactors I-IV) shows experimental set-ups for different occasions. Reactors I-II are supposed to be used for evaluation of a single passage of all cells in an area of interest and then a set-up for a repetition of the same conditions. This is to reveal any metabolic change caused by altered environmental conditions and to see if repeated passages change this response although the conditions are the same. It has been observed in earlier studies some microbial model systems, e.g. the mixed acid fermentation of E.coli, showed this behaviour (Cleland et al, 1990).

The two set-ups, described in the lower part of the figure, describe the two- compartment reactor systems. These can be looked upon in two ways. As the first assumption, this system is a simple scale down approach of a large scale reactor with the well stirred area around the impeller as a stirred reactor. The second part is either 1) a plug flow reactor, operated at different residence times, resembling the residence time distribution of cells in a segregated part of a large scale reactor, or 2) a stirred reactor. Preferably, these reactors are chosen as ideal as possible to avoid a more complicated system. The presence of the stirred reactors adds their residence time distribution to the interpretation of the results. Here, the plug flow reactor has the advantage of the well defined residence time. The second way of looking at these two compartment reactors are as a well defined experimental set-up for kinetic studies of oscillating behaviour.

The set-ups described above, have therefore different possibilities and drawbacks. They will have to be used in parallel to give a deep insight into the metabolic changes possible during large scale cultivations. However, they have to be chosen according to the specific cellular system and metabolism of interest and on what reactor concept or conditions that is to be studied.

MICROBIAL REACTIONS PREDICTED

In the cells, thousands of reactions occur. Which reactions that will be valid to study during the special situation of the oscillating environment of the cells in a production scale reactor seems as an impossible task. However, the cellular reactions are organised in reaction sequences that rule specific events and this makes things somewhat easier.

In fed batch processes, the chosen limiting substrate is often the energy source. Gradients of this substrate are therefore likely to appear. A change in the substrate concentration will affect the **substrate uptake** system, if such an active system is available. Some cells also have more than one route for uptake. Depending on the concentration, either of these are dominating. Normally, the lower the concentration, the higher affinity for the substrate i.e., the lower the saturation constant. It can be shown by simulation, that the saturation constant for substrate uptake (k_S) is an important parameter for process performance. This is shown in fig. 3 below.

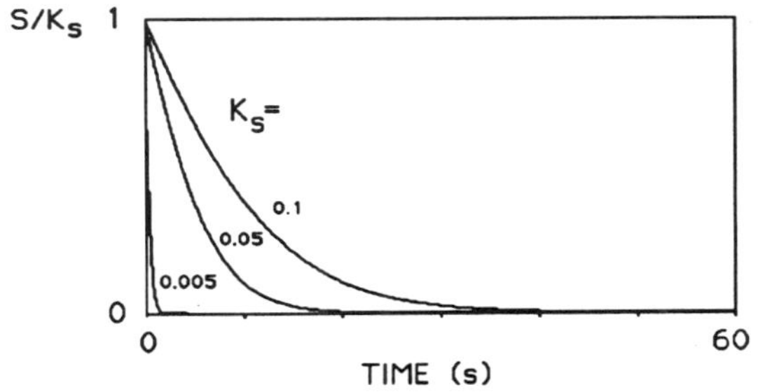

Fig. 3

This figure shows the time before the substrate (dimensionless substrate concentration s/k_s) is completely exhausted in an area of no input of substrate. The figure shows the importance of values which are taken from some typical microorganisms. The biomass concentration in the simulated fed batch process, is 40 g/L in this case. If the substrate uptake rate is affected, the productivity of the process will change.
The level of substrate can also, over a critical value, change the microbial choice of the metabolic pathway. An example of this is the **overflow metabolism** of Saccharomyces cerevisiae, that gives ethanol production if a critical value of approximately 110 mg/L of glucose is exceeded (Enfors et al, 1990). Another example is acetate production by Esherichia coli. This metabolic change will lower yield of biomass produced per sugar consumed.

Gradients of oxygen in bioreactors have been shown a number of times to be severe to the process. These are also very likely because of the low solubility of oxygen in spite of the inhomogenuous mixing conditions. The respiration of Penicillium chrysogenum was irreversible inhibited by short residence times of anaerobic conditions (Larsson and Enfors, 1985 1988). The penicillin production of this organism was changed even before the respiration under cycling oxygen conditions (Vardar and Lilly, 1982).

Other multigene systems are the **stress responses**. These systems have only been investigated to a large extent the last twenty years. The results from these systems are different while their products are to a certain extent the same. The heat shock response (see e.g. Goff et al, 1988) has been described by several authors, and is among other components induced by temperature shifts. The proteins produced by these genes have in some cases been identified as proteases. Other types of stress responses are the stringent response (see e.g. Gallant, 1979) and the starvation response (Groat and Matin, 1986) that shares some of the features of the heat shock response, but still others are response specific. It has been shown that i.e. the heat shock response is induced very rapidly, in the range of mixing times in a large reactor. If proteases are produced, it is a possibility that these might degrade the protein product. Some of the stress proteins are chaperones i.e they shield proteins and take part in the folding mechanism (Skowyra et al, 1990, Goloubinoff et al, 1989). This is of course a positive effect if it could be used, which however remains to be shown.
The list above of different effects on parts of the microbial cellular system is by no means complete and serves as an illustration of complexity of the cellular response and the necessity to investigate the microbial performance during bioprocess calculations. However, in general only some metabolic systems systems of the cells are directly of interest. This is because their time constants of change are in the same range as the time constants of mixing. The examples above belong to this category.

THE MICROBIAL MODEL SYSTEM

The ethanol production of Saccharomyces cerevisiae due to high sugar concentrations is chosen. When the respiratory capacity of the microbial cells is reached, sugar is still taken up by the cells. This overflow is converted to ethanol which can later be consumed by the cells. This depends on the sugar concentration and, passing a critical value of approximately 110 mg/L, ethanol consumption starts (Enfors et al, 1990). However, these cells also produce ethanol as a result of oxygen starvation. To avoid confusion the system was designed to operate under excess of oxygen throughout.

THE SCALE DOWN REACTOR CONSTRUCTION AND SET-UP

In order to investigate to what extent concentration gradients have any influence on the microbial metabolism and to clarify whether or not this can change the bioreactor performance, a scale down reactor was constructed to be used with the microbial model system described above. The objective of this construction was to model a large scale reactor as composed of two different regions, one stirrer region, where most of the energy dissipation occurs, and the rest of the reactor as segregated with a residence time distribution of the liquid flow. Therefore, a two compartment reactor system was designed to consist of a stirred reactor and a plug flow reactor between which the the cells and medium were circulated. The circulation time was set by changing the flow of a circulation pump in front of the plug flow reactor part. This set up is shown in figure 4, below.

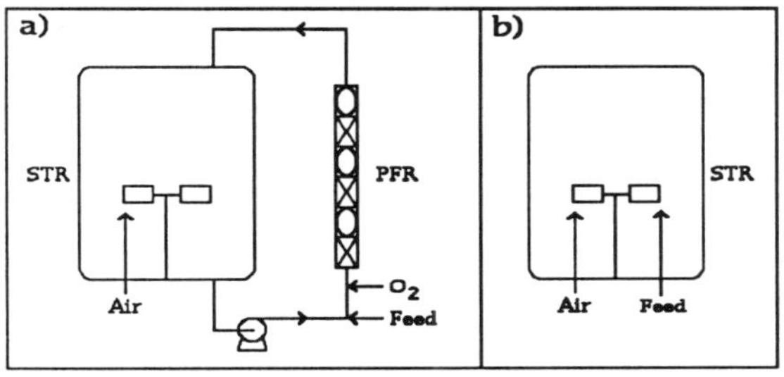

Fig 4

However, both reactors had to be aerated in order to avoid anaerobic ethanol production. The importance of a defined residence time in the plug flow unit and a high oxygen transfer rate led to the construction of an aerated plug flow reactor. The choice was a set up of static mixers. Air or oxygen together with the feed, was introduced perpendicular to the cell suspension in front of this reactor. The reactor was characterised with respect to oxygen transfer and plug flow behaviour. Values of volumetric oxygen transfer coefficient, kLa, were calculated from oxygen transfer rates (OTR), determined by the sulphite oxidation method. Different aeration rates and residence times were compared. The result, in fig 5, shows that kLa values seem more or less independent of the liquid flow rate. The gas flow rate, however, had a clear effect on the oxygen transfer. A maximum of 1130 h^{-1} was found at a rate of 2.2 VVM (volume of gas/liquid volume of reactor, minute).

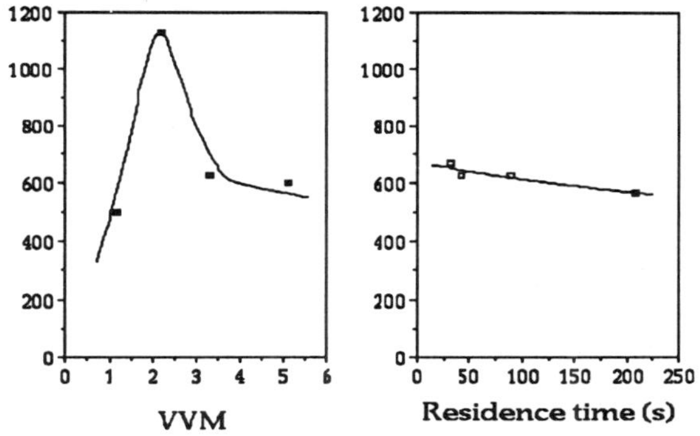

Fig 5 kLa (h^{-1}) as a function of air flow and liquid residence time

Values of kLa obtained from sulphite oxidation are often regarded as overestimated and too high compared to those present in real fermentation media. A difference is of course the properties of coalescence that will be different between the sulphite system and the fermentation medium. Analysis of the dissolved oxygen tension (DOT) at the plug flow outlet during fermentation showed that aerobic conditions sustained up to a biomass concentration of 38 g/L. By calculation, this gives an oxygen transfer coefficient of 340 h-1 for aerobic conditions. Since the sulphite oxidation values were about three times higher, these results show that the values have to be taken with some consideration.

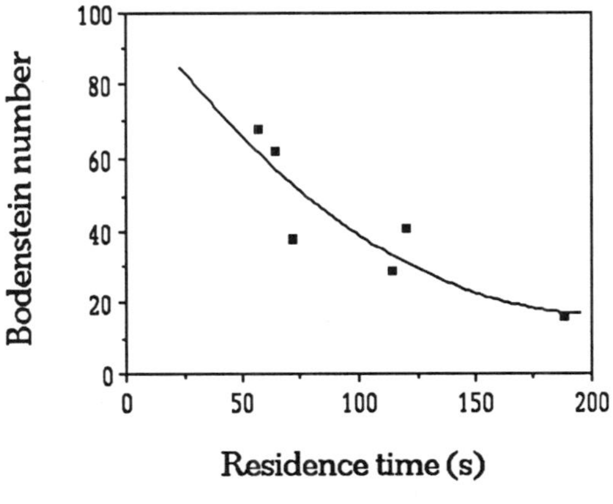

Fig 6 Bodenstein number=$2*t^2/\sigma^2$ where t is the mean residence time and σ is the standard deviation

The plug flow behaviour is represented by the

appropriate Bodenstein number calculated from the results of step change experiments with CBB (Coomassie Brilliant Blue) coloured yeast suspension. A number higher than ten is generally considered as approaching plug flow behaviour. The dependency of the Bodenstein number on the residence time is shown in fig 6. From these studies we conclude that it is possible to have a narrow RTD even when the PFR is aerated. Hence, the effects observed, by introducing the PFR in the scale-down reactor system, should be due to a well defined time exposure of the organism to the PFR environment. By application of different feeding strategies to scale-down reactor cultivations, effects of different concentration gradients can be studied. For example, by feeding the limiting substrate in the STR, the concentration will decline in the PFR, due to the consumption, and eventually become zero. Effects of exposures to low concentrations can then be studied. By feeding at the PFR inlet, the cells will instead repeatedly be exposed to high concentrations of the substrate, as in this investigation.

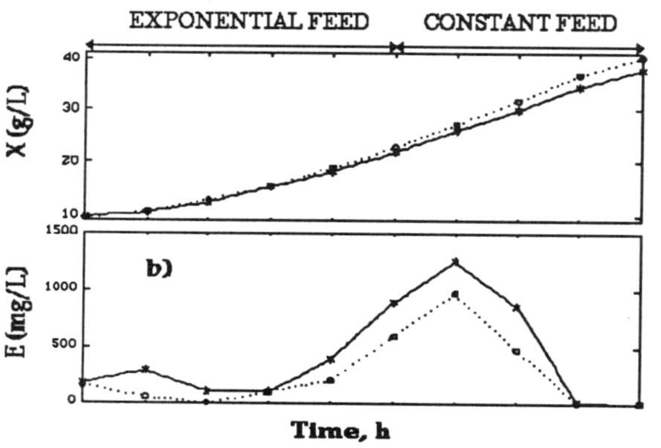

Fig 7 E=ethanol concentration, X=cell mass concentration

The influence on the yield and thereby the performance of a bioreactor, with respect to the Baker's yeast process is shown in fig 7. The dashed lines above shows the results from the two compartment reactor with the feeding position in front of the plug flow reactor. The solid lines are the results from the control reactor that is a stirred well mixed reactor of the same volume as the two compartment system. These reactor set ups and their feeding point location, where earlier shown in fig 4. The result, presented as ethanol and cell mass production is shown in fig 7a and 7b.

CONCLUSION

A number of microbial reactions can have a major importance of the outcome of a bioprocess. By use of the scale-down reactor, hereby simulating the process behaviour, we have shown effects on the yield of biomass in the Baker's yeast process is seen. Whether or not the feed point is positioned in the segregated or the well-mixed part of the reactor is hereby of importance. Feeding in the low micromixing region triggers the overflow metabolism of the yeast, which results in 6 % lower biomass and higher ethanol production correspondingly. This means that the feeding strategy can determine the process performance.

LITERATURE CITED

Cleland N, Larsson G, Enfors S-O; "Characterization of a biological test system for studies of insufficient mixing in bioreactors: H2 evolution from E.coli", Bioproc Eng, 5, (1990)

Enfors S-O, Hedenberg J, Olsson K; "Simulation of the dynamics in the Baker's yeast process", Bioproc Eng, 5, 191-198, (1990)

Gallant J; "Stringent control in E. coli", Ann Rev Genet 13, 393-415, (1979)

Goff S, Voellmy R, Goldberg A; "Protein breakdown and the heat shock response", In: Ubiquitin, chap 8, Ed, Rechsteiner M, Plenum Publishing Corp N.Y. 1988

Goloubinoff P, Christeller J T, Gatenby A A, Lorimer G; "Reconstitution of active dimeric ribulose biphosphate carboxylase from an unfolded state depends on two chaperonin proteins and Mg-ATP", Nature 342, 884-889 (1989)

Goloubinoff P, Gatenby A A, Lorimer G; "GroE heat-shock proteins promote assembly of foreign prokaryotic ribulose biphosphate carboxylase oligomers in Escherichia coli", Nature, 337, 44-47 (1989)

Groat G, Matin A; "Synthesis of unique proteins at the onset of carbon starvation in E.coli", J Ind Microbiol, 1, 69-73, (1986)

Larsson G, Enfors S-O; "Influence of oxygen starvation on the respiratory capacity of Penicillium chrysogenum", Appl Microbiol Biotechnol, 21, 228-233, (1985)

Larsson G, Enfors S-O; "Studies of insufficient mixing in bioreactors: Effects of limiting oxygen concentrations and short term oxygen starvation on Penicillium chrysogenum", Bioproc Eng 3, 123-127, (1988)

Skowyra D, Georgopoulos C, ZyliczM; "The E.coli dnaK gene product, the hsp70 homolog, can reactivate heat inactivated RNA plymerase in an ATP hydrolysis dependent manner", Cell, 62, 939-944 (1990)

Sweere A P G , Luyben K Ch A M, Kossen N F W; "Regime analysis and scale down: Tools to investigate the performance of bioreactors", Enz Microbial Technol 9, 386-398, (1987)

ACKNOWLEDGEMENT

This work was sponsored by the Swedish National Board for Technical Development and by a grant from the Nordic Programme on Bioprocess Engineering under the auspices of NI, the Nordic Fund for Technology and Industrial Development.

We also want to thank Svenska Jästfabriks AB, where parts of this work were carried out.

Geometric Relationships for Scale-Up of Diverse Mixing Processes

James Y. Oldshue

Oldshue Technologies International, 811 Ayrault Rd., Fairport, NY 14450

Large tanks are quite different in many mixing parameters than small tanks used in pilot scale. Geometric similarity controls various mixing parameters in a well-defined fashion and may be suitable for scale-up of some mixing processes. However, if these relationships are not suitable, then consideration should be given to a non-geometric similarity relationship between pilot and full scale. Among the mixing parameters that should be considered are overall blend time, the standard deviation of blend times around the mean, and the variety of shear rates in large tanks compared to small tanks. The importance to the process of shear rate, and particularly if it is important, to the macro and micro scale relationships must be considered as well as the increase in Reynolds number that occurs as tank size increases.

INTRODUCTION

Geometric similarity seems to be a logical requirement for successful scale up. However, it doesn't give a suitable relationship between many important mixing parameters, and often should not be used at all. There are two other types of process similarity that must also be considered. One is kinematic similarity which requires that all the velocities in the various parts of a mixing tank be similar on various scales. This is not normally used in mixing scale up relationships.

Dynamic similarity requires that the ratio of all the forces be similar between different scales. In the case of fluid mixing, it requires that we look at four different kinds of fluid force quantities. One is the inertia force put in by the mixer. The other three are the opposing forces in the process of viscosity, gravity and inter-facial tension.

MACRO/MICRO SCALE RELATIONSHIPS

It is also instructive to look at the way different size process participants react to mixing variables. Looking first at turbulent flow, all of the power consumed by any impeller of any shape and size is converted into heat. This conversion mechanism requires going through viscous shear stresses. The only place in turbulent flow that there is viscous shear stresses are down where what I will refer to as the "micro scale" environment. I am going to define micro scale mixing as beginning about 100 microns, although this is a very soft, qualitative number. In that general area and going downward in size, the process participants would not care much what impeller was used to generate this environment. It is primarily a matter of the energy dissipation at a given site. Energy dissipation is far too simplistic to be the entire story of micro scale mixing, but it is one measure of the micro scale regime that the process participants experience.

At the other end of the scale, we can take measurements of velocities in various parts of the tank and at a given point and characterize the velocity by the sum of an average component (v) plus a fluctuating component, v′ above and below this average. The fluctuating part goes to make up the micro scale environment, and is where the power is ultimately dissipated. By using the average velocity, we can calculate what I am going to call "macro scale" shear rates. Any process participant greater than 1,000 microns will respond to these types of effects. These velocities in the macro scale region are energy transfer agents transferring energy down to the micro scale where it is changed into heat. In this macro scale region, every geometric variable has some effect on the fluid regime, i.e. impeller shape, number of blades, thickness of stock used to make the blades, baffles, tank shape, impeller location and many others.

In viscous flow we have viscous shear stresses at all scales. There is a continual dissipation of energy all the way down from the macro to the micro, Figure 2.

STEP #1 - WHAT TO DO FIRST

My suggestion is that you first ask yourself 'is there any role for fluid shear stresses in determining and obtaining the desired process result?' About half of the time the answer will likely be 'no'. That is the percentage of mixing processes where fluid shear stresses either have no effect or seem to have no effect on the process result. In these cases, mixer design can be based on pumping capacity, blend time, velocities and other matters of that nature. Impeller type, location, and other geometric variables are major factors in these types of processes.

However, if the answer to this first question is 'yes'; there is an effect of fluid shear stresses on the process, then there needs to have a second question asked - 'is it at the micro or macro scale that the process participants are involved' -- and, of course, it may be both.

SCALE UP / SCALE DOWN

Figure 1 shows what happens to many of the variables on scale up. A summary of this is that blend time typically increases and the standard deviation of circulation times around the mean circulation times also normally increases. The quantitative effect depends somewhat on the degree of uniformity required and the blend time being considered.

As a general rule, the operating speed of the mixer tends to go down, while the peripheral speed of the impeller tends to go up. The speed of the mixer is related to the average impeller zone macro scale shear and thus typically goes down in scale up while the impeller peripheral speed is often related to the maximum impeller zone macro scale shear rate, Figure 2. Out in the rest of the tank (away from the impeller) there another spectrum of shear rates which typically is about a factor of 10 lower than the average impeller zone shear rate. These particular impeller zone shear rates tend to decrease on scale up.

The micro scale environment tends to have a power per unit volume of dissipation around the impeller about 100 times higher than it is in the rest of the tank, more or less regardless of the tank size. Thus, the magnitudes of these quantities can be quite similar. This brings up another consideration. paragraph following.

SHEAR RATE MAGNITUDE & TOTAL SHEAR WORK

Shear stresses and their origin from shear rates gives the magnitude of the shear stress environment that the process participants see. The time they are exposed to that magnitude is a major factor in the process result. For example, it may take a minimum shear stress magnitude to create a certain size particle. However, the ultimate distribution of particle sizes may well relate to the length of time that a particle is exposed to that shear rate. The product of shear stress and time determines what is likely to happen to the process. This obviously is a matter of the spectrum of shear stresses throughout the tank and the statistical distribution of circulation times that particles have going through these zones.

With constant viscosity between the model and the prototype and/or a constant change in viscosity to the process during a batch operation, we can substitute shear rate for shear stress and the product of shear rate times the time is dimensionaless number. Considerable progress is being made toward calculating the velocities, shear rates and circulating times in mixing vessels, and suitable models and calculations could be made to model these effects in more quantitative detail both on a point by point basis and at an overall vessel average. What still is challenging, however, is that it is not usually known what effect these particular properties will have on the process participants in a given process and, thus, it is usually necessary to measuring the process result either full scale in the plant, or in smaller size systems in pilot plant or laboratory.

To summarize the situation, geometric similarity controls no mixing variable whatsoever. The question is, 'does that make a difference to the process'. In the portion following, we will take a look at the ten basic mixing technology classifications and see what effect these considerations might have. Added to this is the fact that most industrial mixing processes involve two or more of the ten mixing technological classifications and so their interaction between those technology classification parameters must be considered to give the overall performance of the mixing process.

EFFECT OF CHANGING THE CIRCULATION TIME AND THE SPECTRUM OF SHEAR RATES ON 10 DIFFERENT MIXING TECHNOLOGIES

Gas-Liquid Dispersion. Large tanks tend to have a greater variety of bubble size distribution than do pilot tanks. It is typical that the linear, superficial gas velocity be higher in large tanks than in small tanks which effects dispersion and mass transfer. Gas-liquid systems are very sensitive to the chemical properties of the fluid and the effect of electrolytes, surface active agents and other considerations in the chemistry of the system including viscosity and percent solids effect the relationship.

Gas-Liquid Mass Transfer. If we are concerned with only with the total volumetric mass transfer rate, then we can achieve very similar Kga values in large tanks and in small tanks.

Blend time enters into the picture primarily for other process steps immediately preceding or following the gas-liquid mass transfer step. Blending can play an important role in other steps in the total process in which gas-liquid mass transfer is only one component.

Solids Suspension and Dispersion. For the suspension of solids of a free settling nature up to about 30% by weight, there is not much effect of blend time and shear rate relationships on scale up. However, if a slurry becomes concentrated, on the order of 40% by weight or more, and it some times can go up into the vicinity of 60 or 70 by weight, then the viscosity of the slurry changes the controlling mechanism from suspension of free settling particles to the blending and circulation pattern of a viscous pseudo-plastic material.

Processes using animal cells in biotechnology applications are very sensitive to shear rate since animal cells do not normally have a tough skin like that of fungal cells and are very sensitive to mixing effects. Many approaches have been and are being tried to minimize the effect of increased shear rate on scale up. These include encapsulating the organism in or on micro particles and/or conditioning the animal cells to withstand required shear rates.

Solid-Liquid Mass Transfer. There is potentially a major effect of both shear rate and circulation time in these processes. We look at the slip velocity of the particle, and also whether we can break up agglomerates of particles which may enhance the mass transfer. When the particles become small enough, they tend to follow the flow pattern, so the slip velocity necessary to affect the mass transfer becomes less and less available, Figure 3.

What this shows is that going from the conditions of off-bottom motion to complete uniformity, the effect of mixer power is much less than from going to on-bottom motion to off-bottom suspension. The initial increase in power causes more and more solids to become in active communication with the liquid and has a much greater mass transfer rate than that occurring above the power level for off-bottom suspension in which slip velocity between the particles of fluid is the major contributor.

Since there may well be chemical or biological reactions happening on or in the solid phase, depending upon the size of the process participants, macro or micro scale effects may or may not be appropriate to consider.

In the case of living organisms, their access to dissolved oxygen throughout the tank is of great concern. Large tanks in the fermentation industry often have a Z/T ratio of 2:1 to 4:1, and, thus, top to bottom blending can be a major factor. Some biological particles are facultative and can adapt and re-establish their metabolism at different dissolved oxygen levels. Other organisms are irreversibly destroyed by sufficient exposure to low dissolved oxygen levels.

Liquid-Liquid Emulsions. Almost every shear rate parameter we have effects liquid-liquid emulsion formation. Some of the effects are dependent upon whether the emulsion is both dispersing and coallescessing in the tank, or whether there are sufficient stabilizers present to maintain the smallest droplet size produced for long periods of time. Blend time and the standard deviation of circulation times effects the length of time it takes for a particle to be exposed to the various levels of shear work and thus the time it takes to achieve the ultimate small particle size desired.

As an aside, when a large liquid droplet is broken up by shear stress, it tends to initially elongate into a 'dumbbell' type of shape which determines the particle size of the two large droplets formed. Then, the neck in the center between the 'dumbbell' may explode or shatter. This would give a debris of particle sizes which can be quite different than the two major particles produced.

Liquid-Liquid Extraction. If our main interest is in the total vòlumetric mass transfer between the liquids, the role of shear rate and blend time is relatively minor. However, if we are interested in the bubble size distribution, and we often are because that affects the settling time of an emulsion in a multi-stage co-current or counter-current extraction process, then the change in macro and micro rates on scale up is a major factor. Blend time and circulation time are usually not a major factor on scale up.

Blending. If the blending process is between two or more fluids with relatively low viscosity such that the blending is not effected by fluid shear rates, then the difference in blend time and circulation between small and large tanks is the only factor involved. However, if the blending involves wide disparities in the density of viscosity and surface tension between the various phases, then a certain level of shear rate may be required before blending can proceed to its ultimate degree of uniformity.

The role of viscosity is a major factor in going from the turbulent regime, through the transition region, into the viscous regime and the change in the rule of energy dissipation discussed previously. The role of non-newtonian viscosities comes into the picture very strongly since that tends to markedly change the type of influence of impellers and determines the appropriate geometry that is involved.

Another factor here is that the relative increase in Reynolds number on scale up. This means that we could have pilot plants running in the turbulent region as well as the plant. We could have the pilot plant running in the transition region and the plant in the turbulent, and the pilot plant could be in the viscous region while the plant is in the transition region. There is no apparent way to prevent this Reynolds number change upon scale up in marked way. In reviewing the qualitative flow pattern in a pilot scale system, it should be realized that the flow pattern in the large tank will be at apparently much lower viscosity and therefore at a much higher Reynolds number than is being observed in the pilot plant. This means that the role of tank shape, D/T ratio, baffles, and impeller locations can be based on different criteria in the plant size unit than in the pilot size unit under observation.

Chemical Reactions. Chemical reactions are influenced by the uniformity of concentration both at the feed point and in the rest of the tank and can be markedly affected by the change in overall blend time and circulation time as well as the micro scale environment. It is possible to keep the ratio between the power per unit volume at the impeller and in the rest of the tank relatively similar on scale up, but much detail needs to be considered when talking about the reaction conditions, particularly where it involves selectivity. This means reactions that can take different paths depending upon chemistry and fluid mechanics, and is a major consideration in what should be examined. The method of introducing the reagent stream can be projected in several different ways depending upon the geometry of the impeller and the feed system.

Fluid Motion. Some times the specification is purely in terms of pumping capacity. Obviously, the change in volume and velocity relationships depends upon the size of the two and three-dimensional area or volume involved. The impeller flow is treated in a head/flow concept and the head required for various types of mixing systems can be calculated or estimated.

Heat Transfer. In general, the fluid mechanics of the film on the mixer side of the heat transfer surface is a function of what happens at that surface rather than the fluid mechanics going on around the impeller zone. The impeller provides largely flow across and adjacent to the heat transfer surface and that is the major consideration of the heat transfer result obtained. Many of the correlations are in terms of traditional dimensionaless groups in heat transfer, while the impeller performance is often expressed as the impeller Reynolds number.

WHAT TO DO IN THE PILOT PLANT?

There are several considerations to bear in mind when planning a pilot plant program.

1. The pilot tank is blending rich while full scale tanks are blending poor. This means that relatively inefficient blending impellers are needed in the pilot plant to correspond to the blending efficient impellers used in the plant.

2. One technique to make the pilot plant unit more similar to the plant scale unit is to use impellers of relatively narrow blade width compared to their traditional blade widths used with commercial impellers in the plant. This is purposely reducing the blending performance and improving the shear rate performance in the pilot plant by using impellers of relatively narrow blade width. The blade width cannot be so small that it gets out of proportion to the process participant particles.

3. Always bear in mind the qualitative relationship with viscosity is that the full scale tank will appear to be less viscous than the pilot plant tank, somewhere in the range of a factor of 10 to 50.

4. If it appears that upon a qualitative examination of that the role of circulation time, blend time and shear rate may not be important to the process on scale up, then go ahead and use geometric similarity on the pilot plant study and all of these differences noted above will play no part in the results of the scale up prediction. It may be that there are compensating effects that while circulation time becomes longer and shear rates become larger, there is a compensating effect that makes the process result satisfactory.

SUMMARY

If it is desired to control some particular mixing parameter or several mixing parameters, it may be that a non-geometric similarity approach will give a better representation of full scale performance on the pilot scale.

NOMENCLATURE:

Kla	=	Gas-liquid mass transfer coefficient.
Kga	=	Gas-liquid mass transfer coefficient.
ß	=	Blend time
N	=	Impeller Speed
D	=	Impeller Diameter
Z	=	Liquid level
T	=	Tank diameter
P	=	Power
N_{Re}	=	Reynolds number
N_{NU}	=	Nusselt number
N_P	=	Power number
F_I	=	Inertia Force
$F_{(\mu)}$	=	Viscous Force
$F_{(\sigma)}$	=	Surface tension force
F_g	=	Gravitational Force
μ	=	Viscosity
g	=	Gravity
(ρ)	=	Density

REFERENCES:

1. Oldshue, J.Y., "Mixing '89", <u>Chemical Engineering Progress</u>, 5/89, pp 33-42.
2. Oldshue, J.Y., "Fluid Mixing Technology", McGraw-Hill, 1983.

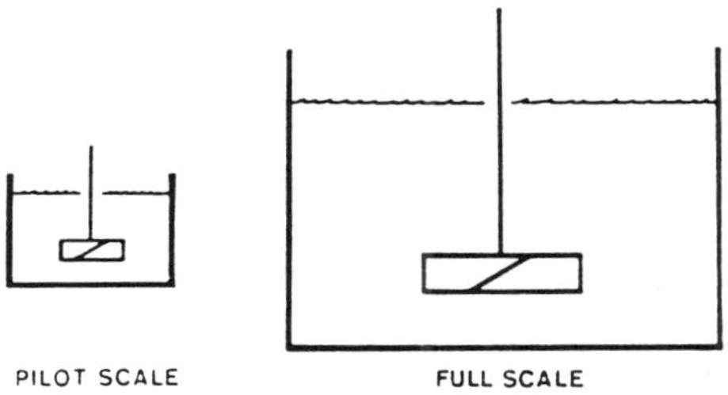

FULL SCALE COMPARED TO PILOT SCALE

BLEND TIME	LONGER
MAX. IMPELLER ZONE SHEAR RATE	HIGHER
AVE. IMPELLER ZONE SHEAR RATE	LOWER
AVE. TANK ZONE SHEAR RATE	LOWER
TURBULENT SHEAR RATES	DIFFERENT

1. Principal changes and mixing parameters on scale up.

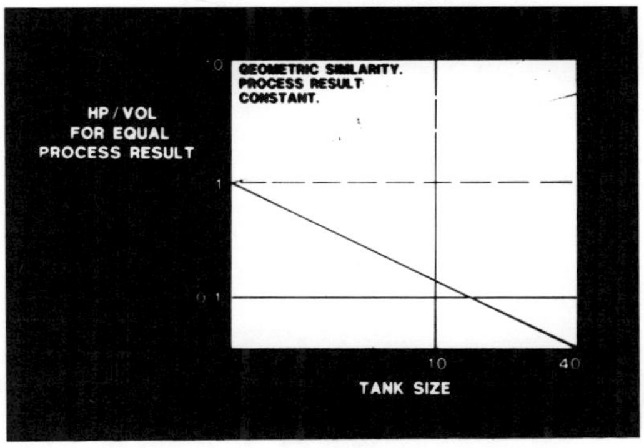

2. Change in maximum impeller zone shear rate and average impeller zone shear rate upon scale up.

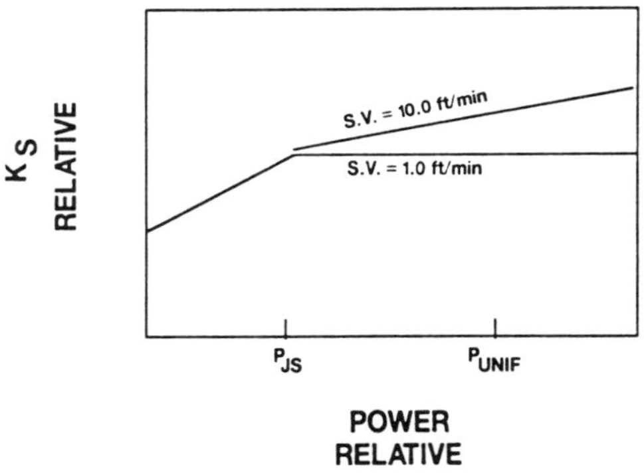

3. Effect of power on liquid-solid mass transfer rate for different solid suspension regimes and different particle sizes.

Scaling Based Upon Process Similarity and Scale Matching Concepts

Gary B. Tatterson

Chem. Eng. Dept. North Carolina A&T State University, Greensboro, NC 27411

The concepts of scaling upon process similarity and scale matching are introduced as alternative to both geometric and non-geometric scale-up methods. A discussion of process objective is given. Scale mismatch is presented as a method to identify potential process failures on scale-up or scale-down. Effects of volume and surface area changes are noted in scaling.

Process Objectives

The most important single item in scaleup is the determination of process objectives. They are what you want the process to accomplish and should be listed in a clear manner. Process objectives have different levels of difficulty and complexity. Process objectives range from general to specific statements and from vague and undefined to detailed statements. Process objectives are easy to accomplish and not so easy to accomplish.

Process objectives determine the type of process and contain information which characterize the process. For example, either a dye or a pigment can be used to obtain colored polymer. Color durability is obtained from pigments since dyes fade. Mixing of dye into polymer is an easy blending operation. The dispersion of pigment into a polymer is a difficult task. Process objectives should state whether color durability is desired.

Process objectives determine equipment. For example, heat transfer applications require good circulation. High shear near the impeller is not needed. High pumping is. High shear and adequate circulation are needed to create emulsions. Multiple objectives may require different impellers.

Process failure indicates the need for another type of equipment, another processing strategy and the need to change process objectives.

Similarity

Single phase flow behavior forms the basis for mixing problems. The basic design step is to establish a good mixing design.

Great similarities between processes exist. For example, liquid mixing times in a gas liquid system are similar to those in single phase. Circulation times for drops are similar to circulation times in single phase mixing. Power draw in a dilute solids system is about the same as in single phase liquid.

Effects of multiphases are often secondary. Behavior of single phase systems is similar to multiphase systems and serves as approximation for multiphase behavior.

Information needed in multiphase is expanded greatly over single phase. There are simply more questions to be answered. Significant similarity exists between multiphase systems. For example, gas dispersion and solids dissolution can be both mass transfer controlled. Both processes depend upon power per unit volume. Multiphase process in a particular geometry provides information concerning another multiphase process in the same geometry.

Static mixer and agitated tank are similar in the formation of an emulsion. Drop breakup is similar in both devices and follow the same correlation (Calabrese, 1992; Calabrese, et. al.)

Stirred Tanks: $d_p/D = 0.054\ We^{-3/5}$

Kenics Mixer: $d_p/D = 0.49\ We^{-3/5}$

Similar looking equipment will behave similarly.

Scale and Process Type

Processes are often surface area, volume or surface-to-volume processes. Substantial changes in volume and surface area per volume occur with scale changes which cause processes to change with scale. Table 1 calculates the relative change in area and volume as a function of scale.

There is much more volume on scaleup than is typically realized. Volume dependent processes are more difficult to represent on the small scale and to scaleup to larger sizes.

Highly exothermic reactions with large heat release can not be accomplished on the large scale because the heat generated can not be removed fast enough.

Power

Power, P, is energy per time added to the process by a number of different mechanisms. Power is an integral quantity which is required to accomplish the process objectives. Power is added to the fluid by the mixing device, e.g., impeller, or by the process itself, e.g., boiling. Power determines other process quantities and utility costs and accomplishes process objectives.

Power, delivered to the process, should be measured and calculated for any mixing and contacting operation. A power analysis of the process should be performed. There is significant uncertainty in power consumed in actual operations.

The Power Analysis

The different mechanisms by which power is added to a process should be identified. Performing a power analysis of the process allows one to weigh the different sources of power and judge their importance. Table 2 lists many of the important sources of power input to a process.

Power input levels vary considerably depending upon mechanisms and conditions. There is no single dominant power input mechanism. There is no dominant P/V mechanism. The determination of the dominant power input mechanism(s) is needed for process understanding.

Process Translation

Process translation is the process of translating operations to larger scale. Process translation is design. It is not necessary to perform exact scaleup. However in the following discussion, scaleup will mean process translation.

The objective in scaling and translation is to determine the chemistry and physics occurring at different scales. Scaling relies heavily on pilot plant studies.

Some systems do not scale and some process phenomena do not happen on the large scale. Some process phenomena do not happen on the small scale.

Equipment selection should be for easy scaleup and experiments should be performed in scaleable systems. Although similar mechanisms will occur in different equipment, equipment, used in experimentation and piloting, should resemble actual plant equipment. Similarity is highly desirable but it is not necessary to have an exact miniature.

Process Similarity and Scaling:

There are many ways that processes are similar as listed in Table 3. However, in this work, process similarity is emphasized. Process similarity exists when processes accomplish the same objectives by the same mechanisms and produce the same product. All other factors are not required.

Obtaining process similarity is the ultimate objective in scaling. Geometric and non-geometric similarity are subsets of process similarity. Exact scaleup is not required and design improvements can be made on scaleup. To establish process similarity, the identification of the controlling mechanism(s) is required.

Successful scaleup depends upon: 1) sufficient information about controlling mechanisms, 2) clearly stated process objectives, 3) appropriate correlations, and 4) scaling procedures.

Scale Matching

Jet mixers, agitated tanks, in-line mixers,

mills, etc., have similar and dissimilar behavior. Matching important process parameters - time scales, velocity scales and characteristics between scales, help to obtain the proper equipment for the different process scales.

Scale matches must be made between:
- Equipment and Process in Selecting Equipment
- Process and Process in Scaling
- Equipment and Equipment in Scaling

Great similarity and commonality exist in equipment. Scale matching is very important for process similarity.

If scale matching can not be used, then scale mis-matching can be used to advantage by eliminating poor designs. Scale mis-matches identify potential process failures on scaleup or scaledown. Slow reactions in a fast mixer and fast reactions in a slow mixer, both scale mis-matches, are indications of poor design.

Processes as Determined by Scale

Processes are controlled by length, area or volume. Examples of volume dependent processes are heat removal, cooking, or chemical reactions. Surface area processes can be volume processes. The problem on scaleup is the relative surface area to volume ratio decreases.

Area dependent processes include interfacial mass & heat transfer and surface dependent chemical reactions. Examples of length dependent processes are solid submergence or suspension. Most processes in agitated tanks are area and volume processes since that is why tanks are used in the first place.

The ratio, volume of a reaction/volume of reactor, has no units and does not appear in dimensional analysis. Wall surface area per volume has a much greater effect on the small scale than on the large scale since the unit is inverse length. It is easier to heat or cool a small tank than a large tank. Wall effects and end effects are more prevalent on the small scale. Controlling factors, i.e., surface area, volume, number, etc., change in relative importance with scale.

Problem areas include no clearly defined controlling mechanism and mechanisms which change with scale.

Process Scaling Objectives/Scaling Criteria

What do you want the scaled process to do? Examples include: a) "I need to scale the cooking rate to be the same between the small scale and large scale," b) "I need the mass transfer rate to be the same between the small scale and large scale," c) "I need the interfacial area per volume to be the same between the small scale and large scale."

Such statements as these have to be translated into engineering equations which form the basis of the process translation.

Elementary Scaleup Procedures for Agitated Tanks

Scaleup and process translation require identification of what does the mixing. Mixing mechanisms include:
- o Mechanical Agitation
- o Jet Mixing
- o Gas sparging in liquids
- o Boiling
- o Static Mixers or Pipeline Mixers
- o Sprays
- o Falling liquid films
- o Shaking as in shaker tables
- o Natural convection currents

Scaleup of a mixing dependent process has to be based upon the dominant mixing mechanism(s).

Impeller Dominant Mixing in Agitated Tanks

The power, P, for turbulent flow is:

$$P = N_p \rho N^3 D^5$$

Assuming equal power numbers and density between two scales, 1 and 2, the following hold:

$$\frac{(P/V)_1}{(P/V)_2} = \left(\frac{N_1}{N_2}\right)^3 \left(\frac{D_1}{D_2}\right)^2 \text{ for turbulent flow}$$

A similar relationship can be developed for laminar mixing.

P/V is the amount of energy/time per volume needed to accomplish the process result.

Scaleup Objective: Equal Heat Transfer/Volume

This scale objective is similar to process cooking. Heat, Q, added per volume, V, is proportional to the heat transfer coefficient, h, and area, A, or

$$Q/V \propto hA/V$$

Correlations for the turbulent heat transfer coefficient reduce to:

$$hD \propto (ND^2)^{2/3}$$

Assuming the area for heat transfer, A, scales proportional to D^2 as in a heating jacket

$$Q/V \propto N^{2/3}D^{-2/3}$$

For equal Q/V on scaleup from scale 1 to scale 2, the ratioing of the above equation provides:

$$\left(\frac{N_1}{N_2}\right) = \left(\frac{D_1}{D_2}\right)$$

Given D_1, D_2, and N_1, N_2 is calculated. On a power per volume basis, scaleup follows:

$$\frac{(P/V)_1}{(P/V)_2} = \left(\frac{D_1}{D_2}\right)^5$$

which is impractical. Attempting to maintain equal heating per volume on scaleup becomes very difficult.

The assumption that heat transfer area is proportional to D^2 was a restriction imposed by geometric similarity. If geometric similarity is relaxed, then the scaleup may be practical. If massive tank internals are added and assuming $A \propto D^3$, then

$$Q/V \propto N^{2/3}D^{1/3}$$

For equal Q/V on scaleup

$$\left(\frac{N_1}{N_2}\right)^3 = \left(\frac{D_2}{D_1}\right)^{3/2}$$

and

$$\frac{(P/V)_1}{(P/V)_2} = \left(\frac{D_1}{D_2}\right)^{1/2}$$

which is more practical and explains why massive internals are added on scaleup.

Scaleup Objective: Equal Heat Transfer Coefficients

The process scaling objective of equal heat transfer coefficients on scaleup may mean keeping material from burning or freezing. Correlations for the turbulent heat transfer coefficient reduce to:

$$hD \propto (ND^2)^{2/3}$$

For equal heat transfer coefficient on scaleup:

$$\left(\frac{N_1}{N_2}\right)^3 = \left(\frac{D_1}{D_2}\right)^{3/2}$$

or:

$$\frac{(P/V)_1}{(P/V)_2} = \left(\frac{D_1}{D_2}\right)^{1/2}$$

which is practical.

The above are several scaleup procedures for turbulent impeller dominant mixing in agitated tanks. Many other scaleup procedures are available or can be developed.

Non-geometric Scaleup

Geometric scaleup is not require and is rarely done in practice. Non-geometric scaleup is possible and can be accomplished. All that is required is an understanding of the process objectives and their dependency on geometry. Often, process design improvements can be made when the restriction of geometric similarity is removed.

Conclusions

Scaleup is open ended and is not a clear cut process. Uncertainties are caused by poorly defined process objectives, poor understanding of the process mechanisms and poor understanding on how controlling mechanisms change with scale. The lack of information is most harmful.

Nomenclature

D	impeller diameter
d_p	drop diameter
h	heat transfer coefficient
N	impeller rotational speed
N_P	power number
Q	heat
P	power
V	volume
We	geometry dependent Weber number
ρ	density

Subscripts

1,2	scale 1 and scale 2

References:

Berkman, P.D., and Calabrese, R.V., AIChEJ, 34, 602, 1988.

Calabrese, R.V., Chang, T.P.K., and Dang, P.T., AIChEJ, 32, 657, 1986.

Calabrese, R.V., Wang, C.Y., and Bryner, N.P., AIChEJ, 32, 677, 1986.

Calabrese, R.V., Private Communication, 1992

Tatterson, G.B., Scaleup and Design for Industrial Mixing Processes, McGraw-Hill Publishing Co, in press, (1993)

Table 1

Scale	Length	Surface Area	Volume	Area/Volume	Volume/Area
1	0.01m	3.93×10^{-4} m^2	7.85×10^{-7} m^3	500 m^{-1}	0.002 m
10	0.1m	3.93×10^{-2} m^2	7.85×10^{-4} m^3	50 m^{-1}	0.02 m
100	1.0m	3.96 m^2	7.85×10^{-1} m^3	5 m^{-1}	0.2 m
1000	10m	393 m^2	785. m^3	0.5m^{-1}	2.0 m

Table 2: Different Forms of Power Input

- o Impeller
- o Gas sparging in liquids
- o Boiling
- o Sprays
- o Falling liquid films
- o Shaking as in shaker tables
- o Natural convection currents due to density differences
- o Natural convection because of a chemical reaction

Table 3: Different Types of Similarity

General Similarity
    Process Similarity
Standard Similarity Types in Mixing and Fluid Mechanics
    Geometric (Same Geometric Shape)
    Kinematic (Same Velocity Ratios)
    Dynamic (Same Force Ratios)
Less Common Types of Similarity
    Thermal (Same Temperature Gradients)
    Chemical (Same Concentration Gradients)
    Mathematical (Same Numbers and Structure)
Other Types of Similarity
    Color
    Texture

Table 4: PROBLEM AREAS IN SCALING

- o Absolute mathematical equivalence may not be scaleable.
- o Boundary conditions and/or initial contacting conditions change on scaling. (Some boundary conditions are highly surface dependent. How material is added determines the flow environment.)
- o Surface area/volume and/or volume/volume change on scaling.
- o Processing conditions change on scaling.
- o Terms in the fundamental equations change on scaling.
- o Flow regimes change on scaling.
- o Scaling into transition regimes may occur.
- o Larger differences between average and local values may occur in a process.
- o Problems in Control of the Process.

Index